Stainless Steels

Edited by Ambrish Singh

Published in London, United Kingdom

IntechOpen

Supporting open minds since 2005

Stainless Steels
http://dx.doi.org/10.5772/intechopen.96827
Edited by Ambrish Singh

Contributors
Temitope Olumide Olugbade, Maruthasalam Sowrirajan, Selvaraj Vijayan, Munusamy Arulraj, André Paulo Tschiptschin, Carlos Eduardo Pinedo, Selma Attabi, Abdelaziz Himour, Lakhdar Laouar, Amir Motallebzadeh, Swarup Kumar Ghosh, Indrajit Dey, Pallabi Manna, Muralidhar Yadav, Nisith Kumar Tewary, Jayanta Kumar Saha, Tomomi Shiratori, Tatsuhiko Aizawa, Tomoaki Yoshino, Yohei Suzuki, Takafumi Komatsu, Gennadii Snizhnoi, Duduzile Zamavezi Nkomo, Nomsombuluko Masia, Ambrish Singh

First published in London, United Kingdom, 2022 by IntechOpen
IntechOpen is the global imprint of INTECHOPEN LIMITED, registered in England and Wales, registration number: 11086078, 5 Princes Gate Court, London, SW7 2QJ, United Kingdom
Printed in Croatia

British Library Cataloguing-in-Publication Data
A catalogue record for this book is available from the British Library

Additional hard and PDF copies can be obtained from orders@intechopen.com

Stainless Steels
Edited by Ambrish Singh
p. cm.
Print ISBN 978-1-80355-132-6
Online ISBN 978-1-80355-133-3
eBook (PDF) ISBN 978-1-80355-134-0

Meet the editor

Dr. Ambrish Singh is a professor in the School of New Energy and Materials, Southwest Petroleum University, China. He leads the Corrosion Research Group (CRG) with master's and Ph.D. students and post-doctoral researchers. His research interests include corrosion, electrochemistry, green chemistry, quantum chemistry, smart coatings, polymers, nanomaterials, composites, and petroleum engineering. He is the recipient of several awards, including the prestigious Sichuan 1000 Talent Award from the Sichuan Government, China, the President's Award for exceptional post-doctoral research work, and the Young Scientist Award from UPCST(Council of Science and Technology, Uttar Pradesh), Lucknow, India. Before moving to China, Dr. Singh worked as an assistant professor in the Department of Chemistry, Lovely Professional University, India. He has published more than 150 SCI peer-reviewed research papers in high-impact journals. He is a reviewer, editor, and board member for more than forty journals. Dr. Singh has been invited to present his work at several national and international conferences, seminars, and workshops. He received an award for the best oral presentation at the CMSE(Conference on Materials Science and Engineering) international conference. He has drafted and filed five patents. He has completed several state and provincial projects in China and India. He is also a consultant for a project on corrosion inhibitors for the CO2 environment in Saudi Arabia. Dr. Singh is a member of the National Association for Corrosion Engineers (NACE), Society of Petroleum Engineers (SPE), and American Chemical Society (ACS).

Contents

Preface

The book covers some important topics regarding stainless steel corrosion in different corrosive media.

Corrosion, while silent and often subtle, is probably the most significant cause of the degradation of society's structures. Enhancing the life of structures and engineered materials while protecting the environment and public safety is one of the paramount technological challenges for our nation and the world. Corrosion-related problems span a wide spectrum of materials and systems that impact our daily lives. Corrosion seriously affects many sectors of nations' economies. Billions of dollars per year are lost due to corrosion. The cost of corrosion can be reduced to 20% to 25% of this annual figure by applying corrosion control technologies. These include material selection, proper designing, electrochemical protection coating, and inhibitors. Among these methods, inhibitors are used for a wide range of applications in oil industries, pipelines, refineries, domestic central heating systems, industrial cooling systems, pickling of metals, and acidization of oil wells.

Corrosion contributes to losses of 7.5 billion USD in world revenue every year, which is estimated to be 3.4% of the world's total GDP. However, in India, economic loss due to corrosion is 1670 billion USD, which is around 4.2% of GDP. The biggest problem in the industry is corrosion, and various steps have been taken to resolve this problem. In the petroleum industry, stainless steel alloys are commonly used in production, refinery, turbine parts, fasteners, and so on due to their resistance to corrosion as well as high mechanical strength.

Stainless steels consist of chromium, which is resistant to corrosion and heat. The oxide layer appearing on the surface of stainless steel in contact with air and water represents a barrier that protects the steel against undesirable and destructive reactions in surrounding environments. Generally, the corrosion resistance of stainless steel depends on the stability of the oxide film grown on its surface. However, the oxide film or passive film can be altered by severe working conditions like high acidity, high temperature, and the presence of impurities in industrial acids. To overcome this problem, the use of mitigators, coatings, and other protection methods is an effective tool against corrosion. In addition, the use of inhibitors for commonly used stainless steel spares the industry the economic burden of using materials with high levels of alloying elements that are too expensive. Furthermore, as guidelines for the use of inhibitors become more stringent and exigent on the ecological aspect, the development of biodegradable and environmentally compatible or ecofriendly inhibitors is becoming an important issue. The use of polymers, nanocomposites, biopolymers, plant extracts, and many more methods are employed to overcome the corrosion of stainless steel.

I would like to thank all the authors and reviewers for their invaluable contributions.

Dr. Ambrish Singh
School of New Energy and Materials,
Southwest Petroleum University,
Chengdu, China

Section 1

Introduction

Chapter 1

Introductory Chapter: Protection of Stainless Steels in Corrosive Media

Ambrish Singh

1. Introduction

Corrosion has now become the part and parcel of metals and alloys in reactive environments. It is an interdisciplinary action that takes place in various disciplines. It includes chemistry, physics, mechanical engineering, electrical engineering, civil engineering, metallurgy, materials science, new energy devices, electronics, marine engineering, dentistry, bio-implants, aviation, military, and biology. Several mechanisms of action are involved depending on the harsh media and metal surface. Since corrosion is omnipresent wherever metals are present, it causes significant loss of materials and hampers the economy globally. Almost all the industries have to spend a lot of money for repairing, replacement, and supervision of corrosion worldwide. It has to be mitigated in order to run the industries smoothly and to reduce the shutdown and failures. There are numerous methods to mitigate corrosion, but still only 20–30% can be mitigated. So, there is always a motivation and urge to develop new materials or to use prevalent corrosion control techniques to mitigate corrosion. Use of inhibitors, material selection, suitable design, coatings, cathodic/anodic protection, and electrochemical protection are some of them used extensively [1–5].

Among these available techniques, application of inhibitors is easy to apply, and cheap and environmentally benign. The corrosion rate is reduced by the presence of corrosion inhibitors. Corrosion inhibitors incorporate themselves to corrosion product films in such way so as to increase the film's capacity to prevent corrosion. According to NACE, an inhibitor is defined as "a substance which retards corrosion when added to an environment in small concentration" [NACE (1965)] and recent ISO definition of an inhibitor is "a chemical substance which decreases the corrosion rate when present in the corrosion system at a suitable concentration without significantly changing the concentration of any other corrosive agent" [ISO (1986)]. Inhibitors may also be defined on electrochemical basis as substances that reduce the rates of either or both of partial anodic oxidation and cathodic reduction reaction. The corrosive media can be the acidic solutions, alkaline solutions, or the neutral media. Apart from them, gases such as carbon dioxide (CO_2) and hydrogen sulfide (H_2S) present in the solutions can make it more corrosive for the metals. Several organic inhibitors are being used, but due to the regulations and policies, they cannot be applied in higher concentrations to mitigate corrosion. So, as a replacement several organic/inorganic inhibitors were developed by means of the green principles in order to get the less toxic inhibitors that can be used in higher concentrations. Several degradable bio-polymers, plant extracts, ionic liquids, bio-macromolecules, proteins, and drugs were used by different research teams all over the world [6–10].

Stainless steels are used in industries due to their good resistance toward corrosion and high mechanical strength. Several joints and parts are made up of stainless steels in order to reduce corrosion. Iron (Fe), chromium (Cr), and nickel (Ni) are the base elements that are present in the stainless steels. As soon as the stainless steel is exposed to a corrosive media, a thin layer of chromium oxide is formed on the surface of stainless steel due to passivation. The thickness of the film can vary from metal to alloy and from acidic to alkaline medium. The presence of this oxide layer on the metal surface keeps it intact from the reactive environment, and therefore, the steel remains protected from corrosion. The layer is inactive and will not react with any harsh environments being self-repairing in nature. The ability of stainless steel and other alloys to resist corrosion is increased through the process of passivation, which is a chemical treatment. Passivation can also be applied to other metals. There are various advantages of using passivated equipment and systems, including the following: Passivation removes surface contaminants. The process of passivation makes materials more resistant to corrosion. Any steel with 10.5% of chromium and 1.2% of carbon can be termed as stainless steel. The stainless steel can be classified into five types: a) austenitic, b) ferritic, c) martensitic, d) duplex, and e) precipitation hardening. Out of these five steels, the first two (austenitic and ferritic) account for almost 95% of the total steel applications. And, among these two austenitic steels account for 75% of the applications in the market. Stainless steels consist of chromium and widely due to its corrosion resistant and heat-resistant properties [11].

The stainless steel consists of various compositions with iron (Fe) as a base element. It is combined with chromium (Cr) and nickel (Ni) to produce 304 stainless steels. Due to the presence of harsh environments and dynamic conditions, these steels can still undergo corrosion. It is easily affected by the pH, temperature, flow velocity, pressure/load, heat treatment, welding, and stress. Little changes in design, mechanical strength, and resistance properties can help in mitigation of corrosion. If sulfur (S) or selenium (Se) is added, then 303 SS or 303 Se SS can be prepared. If Cr is increased and Ni is reduced, the duplex stainless steel can be prepared. The addition of copper (Cu), aluminum (Al), and titanium (Ti) can lead to the formation of precipitation hardening stainless steels. The addition of Mn and N in 304 stainless steels can form 201 and 202 type stainless steel. If Cr is lowered in 304 SS, then it can lead to the formation of martensitic 403, 410, and 420 type stainless steels. The addition of molybdenum (Mo) in 304 type SS can form 316 and 317 type stainless steels. So all these modifications and heat treatment can be done to obtain the desired stainless steels for the laboratory and industrial usage. All these can be prepared or purchased following the AISI or any other international standards [12].

The book depicts the different studies done by research groups on the stainless steels. It includes the working mechanisms, chemical reactions, and surface characterizations to produce a clear picture and help the readers to develop a good understanding of the experimental situation. This book provides an overview of problems encountered in stainless steels due to corrosion and their control. Stainless steels are almost an essential component in most of the industries, and the satisfactory working of these systems is affected by corrosion, scale formation, and fouling of equipment. Apart from that the usage of traditional and new corrosion monitoring techniques will help readers to gather all the required information at one place. Most of the compiled work is new and not reported elsewhere. It will serve as a good reference for those interested in corrosion studies of stainless steels in different corrosive solutions. The book is useful for professionals, researchers, graduate, and undergraduate students. Moreover, it would be of immense importance to the professionals undertaking specialized short-term courses in this field.

Author details

Ambrish Singh
Southwest Petroleum University, School of New Energy and Materials,
Chengdu, Sichuan, China

*Address all correspondence to: vishisingh4uall@gmail.com;
drambrishsingh@gmail.com

References

[1] Ambrish Singh KR, Ansari IH, Ali YL, El Ibrahimi B, Bazzi L. Combination of experimental, surface and computational insight into the corrosion inhibition of pyrimidine derivative onto Q235 steel in oilfield acidizing fluid under hydrodynamic condition. Journal of Molecular Liquids. 2022;**353**:118825

[2] Ambrish Singh KR, Ansari AK, Alanazi MAQ, Banerjee P. Biological macromolecule as an eco-friendly high temperature corrosion inhibitor for P110 steel under sweet environment in NACE brine ID196: Experimental and computational approaches. Journal of Molecular Liquids. 2022;**345**:117866

[3] Singh A, Ansari KR, Alanazi AK, Quraishi MA, Ali IH, Lin Y. Probing inhibition effect of novel biopolymer-based composite for the inhibition of P110 steel corrosion in 15% HCl under dynamic condition. Sustainable Chemistry and Pharmacy. 2022;**26**:100599

[4] Singh S, Singh R, Sharma NR, Singh A. Extract from Clarias batrachus fins as environmental benign corrosion inhibitor for mild steel in acidic solution. International Journal of Electrochemical Science. 2022;**17**:220341. DOI: 10.20964/2022.03.31

[5] Singh A, Bedi P, Ansari KR, Pramanik T, Chaudhary D, Santra S, et al. Inhibition effect of newly synthesized benzoxanthones derivative on hydrogen evolution and Q235 steel corrosion in 15% HCl under hydrodynamic condition: Combination of experimental, surface and computational study. International Journal of Hydrogen Energy. 2021;**46**:37995-38007

[6] Ambrish Singh KR, Ansari Priyabrata Banerjee, Manilal Murmu MA. Quraishi, Yuanhua Lin. Corrosion inhibition behavior of piperidinium based ionic liquids on Q235 steel in hydrochloric acid solution: Experimental, density functional theory and molecular dynamics study. Colloids and Surfaces A: Physicochemical and Engineering Aspects. 2021;**623**:126708

[7] Ambrish Singh KR, Ansari MAQ, Kaya S, Erkan S. Chemically modified guar gum and ethyl acrylate composite as a new corrosion inhibitor for reduction in hydrogen evolution and tubular steel corrosion protection in acidic environment. International Journal of Hydrogen Energy. 2021;**46**:9452-9465

[8] Singh S, Singh R, Sharma NR, Singh A. Ethanolic extract of cockroach wing powder as corrosion inhibitor for N80 steel in an ASTM D1141-98(2013) standard artificial seawater solution. International Journal of Electrochemical Science. 2021;**16**:210841. DOI: 10.20964/2021.08.43

[9] Ambrish Singh KR, Ansari DS, Chauhan MA, Quraishi HL, Chung I-M. Comprehensive investigation of steel corrosion inhibition at macro/micro level by ecofriendly green corrosion inhibitor in 15% HCl medium. Journal of Colloid and Interface Science. 2020;**560**:225-236

[10] Ambrish Singh KR, Ansari MAQ, Banerjee P. Corrosion inhibition and adsorption of imidazolium based ionic liquid over P110 steel surface in 15% HCl under static and dynamic conditions: Experimental, surface and theoretical analysis. Journal of Molecular Liquids. 2021;**323**:114608

[11] Saleh TA, Haruna K, Nur MM, Alharbi B. Synthesis of amine grafted poly (acrylic-maleic) as an efficient inhibitor against stainless steel corrosion

in a highly saline medium. Progress in Organic Coatings. 2022;**170**:106974

[12] Asfia MP, Rezaei M. A study on localized corrosion behavior of 304 stainless steel in the presence of allium Sativum extract inhibitor using electrochemical noise analysis. Materials Chemistry and Physics. 2021;**274**:125158

Section 2

Corrosion

Chapter 2

Corrosion Resistance, Evaluation Methods, and Surface Treatments of Stainless Steels

Temitope Olumide Olugbade

Abstract

Stainless steels are widely recognized and find applications in many engineering industries and companies due to their excellent properties including high resistance to corrosion as a result of their minimum 10.5% chromium content, exceptional strength and durability, temperature resistance, high recyclability, and easy formability. In the present book chapter, the basic concepts of stainless steel including its applications, classifications, and corrosion properties will first be discussed. Thereafter, their corrosion behaviour will then be explained. The various methods by which the corrosion resistance behaviour can be significantly improved including surface treatments such as coatings/electrodepositions, alloying, mechanical treatment, and others will be discussed in detail.

Keywords: stainless steels, surface treatment, corrosion, passivation, chromium content, electrodeposition

1. Introduction

Stainless steels (iron-based alloys) are widely recognized due to their high machinability, hardness, mechanical strength, good heat resistance and excellent corrosion resistance [1–3]. Compared to other steels, the superior corrosion resistance exhibited by many stainless steels can be attributed to the chromium content (about 10.5 wt.%) which initiated the formation of a stable layer of chromium oxide on the steel surface [2, 3] thereby preventing chemical reactions with the bulk material hence reducing corrosion attack to the minimum. Chromium is one of the major elements that play a vital role in the corrosion resistance of stainless steels. However, when exposed to water for a long period, their corrosion resistances reduce at elevated temperatures, hence the need to come up with more robust techniques for protecting the sample surface.

Even though they exhibit good corrosion resistance, efforts have continually been made to further improve the corrosion resistance behaviour of stainless steels. Heat treatment such as annealing due to the oxidation of steel surface [4–6], surface mechanical treatment [7–12], coatings/electrodepositions [13–16], alloying, machining/molding [17] and many more are presently in use to further improve the corrosion resistance behaviour of stainless steels. Other protection methods include epoxy coating, cathodic protection, and thicker concrete cover. For instance, surface modifications, as well as heat treatments of the modified sample surface

 IntechOpen

by low temperature annealing, were used to enhance the corrosion resistance of 301, 17-4PH, 304 steels, 316, and mild steels [4, 5, 9, 12]. To sum it up, stability, compaction, chemical composition, thickness, and many more are the main factors influencing the corrosion resistance of stainless steels [18–20].

In the present study, the general overview of stainless steels including their properties and application is presented. The corrosion resistance of stainless steel and evaluation methods are evaluated. The surface treatment methods aimed at enhancing the overall corrosion and mechanical properties are then presented.

2. Corrosion resistance of stainless steel and evaluation methods

Due to the nature and change in the environment, many metallic materials are expected to possess a good corrosion resistance against corrosion attacks over time. However, the corrosion resistance ability of materials differs from each other, and corrosion does set in when the corrosion-resistant limit of a material is exceeded [21–23]. Hence, the major reason why many material scientists and corrosion experts always pay much attention on how to continuously protect the material surface from degradation and corrosion via surface treatment method, coatings, and other related techniques. For clarity's sake, the corrosion behaviour can be studied when the material is exposed to an aggressive corrosive environment alone (polarization) [9, 16] or under the action of both tensile stress and corrosion reaction (stress corrosion cracking—SCC) [18].

The conventional polarization tests are normally carried out using an electrochemical workstation consisting of the traditional three-electrode system; (1) reference electrode (RE), whose material can be made of saturated calomel electrode (SCE) or silver/silver chloride (Ag/AgCl), (2) counter electrode (CE), which can be Platinum (Pt), graphite, gold or carbon rod, and (3) working electrode (the testing material).

Generally, the corrosion resistance of metallic materials can be evaluated through electrochemical tests which can be done in the following ways (**Figure 1(a–f)**); (1) open circuit potential (OCP) study, (2) potentiodynamic polarization study, (3) potentiostatic polarization study including the current-time transient (CTT) study and double-log plot, (4) electrochemical impedance spectroscopy (EIS) analysis including the Nyquist plot, Bode impedance, and phase angle plots, and (5) Mott-Schottky analysis which is normally carried out to determine the semiconducting characteristics of the passive film.

To a large extent, the OCP test determines the stability of samples in the electrolyte before performing polarization and EIS tests. Here, it is believed that the higher the corrosion potential, the more stable the sample, and probably the better the corrosion resistance [5, 11], i.e., the sample "A" in **Figure 1(a)** possessed higher corrosion potential and is therefore expected to be more stable than sample "B". The potentiodynamic polarization shows the corrosion behaviour in terms of corrosion current density (i_{corr}) and corrosion potential (E_{corr}) which can be determined from the corrosion graph using the Tafel extrapolation method. It is generally believed that the lower the i_{corr} and higher the E_{corr}, the more the formation of the passive film, hence the better the corrosion resistance [7–9].

In addition, the anodic polarization process can be categorized into four regions, as illustrated in **Figure 1(b)**; (1) activation zone, where the i_{corr} gradually increases with E_{corr}, (2) activation-passivation transition zone, where the i_{corr} decreases gradually and started forming passivation film, (3) passivation zone, which involves further decrease in i_{corr}, signifying the formation of more passivation film, and (4) transpassive zone, signifying the degradation of the passive film with

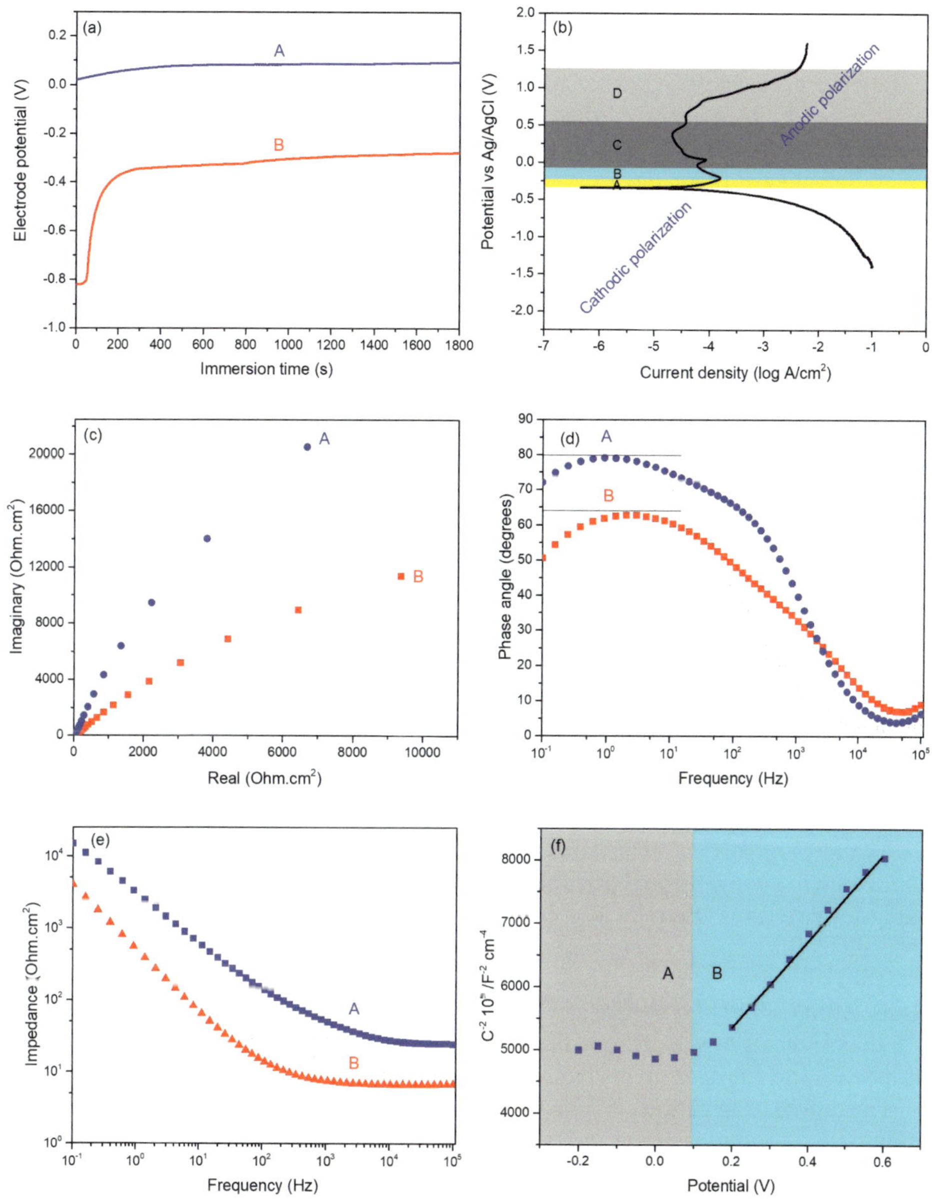

Figure 1.
Illustration of the corrosion properties of metallic materials; (a) open circuit potential (OCP), (b) potentiodynamic polarization, (c) Nyquist plot, (d) Bode phase angle plot, (e) Bode impedance plot, and (f) Mott-Schottky plot.

a rapid increase in i_{corr}. In EIS analysis, the samples with better charge-transfer resistance, higher impedance, and phase angle are believed to have a stabilized and more protecting passive film, hence possessing better corrosion resistance [10, 12]. For instance, sample "A" in **Figure 1(c–e)** is better than sample "B" in terms of corrosion resistance since it has a larger diameter of the semi-circle, higher phase angle, and impedance.

Corrosion fatigue is a common phenomenon that frequently occurs when materials are often exposed to simultaneous actions of corrosive environment and repeated stress, which leads to a markedly decrease in fatigue strength. In addition, unlike stress corrosion cracking which causes intergranular cracking and mostly occurs in harmful environments, corrosion fatigue which causes transgranular cracking, can occur at any time and cannot be avoided in some cases [4, 5, 12].

Furthermore, Mott-Schottky analysis determines the electronic properties of the passive film by measuring the capacitance as a function of potential, which ultimately determines the semiconducting characteristics of the passive film. It is generally believed that a negative slope represents a p-type semiconductor while a positive slope signifies an n-type semiconductor. When a positive slope is obtained, it means there is no change in the semiconducting characteristics of the passive film, hence an enhancement in the stability of the passive film which can eventually increase the corrosion resistance. As illustrated in **Figure 1(f)**, the section denoted by "A" represents the flat band potential zone while the portion denoted by "B" represents the n-type semiconductivity zone.

3. Surface treatments: enhancing the overall corrosion and mechanical properties

The overall corrosion and mechanical properties of stainless steels can be enhanced by subjecting them to different types of surface treatments including anodization, electrodeposition, dip-coating, micro-arc oxidation (MAO), and surface mechanical treatments. Anodization is a surface modification technique involving the passage of constant current and voltage through the cathode and anode resulting in the deposition of the natural oxide layers on the material surface [24] with an improved thickness and properties thereby turning the metal surface into an excellent finish which is anodic oxide in nature, corrosion-resistant, durable, and wear/abrasion-resistant.

As a form of surface treatments (anticorrosion processing methods), electrodeposition is a conventional phenomenon, and the combination of reduction and oxidation reactions whereby dissolved metals and alloys cations are cathodically reduced by the passage of electric current in electrolytes leading to the formation of a thin layer coating on electrodes [25]. By this, thin layers of functional materials including alloys and metals can be electrolytically deposited on the surface of Mg alloys (acting as the cathode in the electrolytic cell) to improve the corrosion properties and the overall mechanical behaviour.

The dip-coating process can be highlighted in five ways depending on the immersion time and speed as well as the withdrawal speed, (a) dipping the substrate into the desired solution, (b) removal of the dipped substrate from the solution, (c) deposition of the film on the substrate after removal, (d) removal of excess liquid from the material surface, and (e) dispersal of the solvent from the liquid film. Meanwhile, micro-arc oxidation is the electrochemical oxidation process through which hard, dense, and protective ceramic oxide coatings are formed on metal surfaces for corrosion protection under the influence of various processing conditions and parameters [26–28].

4. Conclusion

Despite their good corrosion resistance, stainless steels still experience failure, especially when exposed to an aggressive environment for a prolonged period of time. This can be corrected by adopting the right surface treatment methods including alloying, heat treatment, coating/electrodeposition, and surface mechanical treatment. The corrosion resistance of stainless steels often depends on the stability, compaction, chemical composition, thickness of the passive film generated on the material surface.

DOI: http://dx.doi.org/10.5772/intechopen.101430

Author details

Temitope Olumide Olugbade
Department of Industrial and Production Engineering, Federal University of Technology, Akure, Ondo State, Nigeria

*Address all correspondence to: tkolugbade@futa.edu.ng

References

[1] Dias D, Nakamatsu S, Rovere CAD, Otubo J, Mariano NA. Characterization and corrosion resistance behavior of shape memory stainless steel developed by alternate routes. Metals. 2020;**10**:13

[2] Nishimura T. Crevice corrosion resistance and structure of passive film on Fe–Mn–Si–Cr–Ni steel. ISIJ International. 2016;**56**:654-660

[3] Rovere CAD, Alano JH, Silva R, Nascente PAP, Otubo J, Kuri SE. Influence of alloying elements on the corrosion properties of shape memory stainless steels. Materials Chemistry and Physics. 2012;**133**:668-673

[4] Olugbade TO, Lu J. Characterization of the corrosion of nanostructured 17-4 PH stainless steel by surface mechanical attrition treatment (SMAT). Analytical Letters. 2019;**52**:2454-2471

[5] Olugbade T, Liu C, Lu J. Enhanced passivation layer by Cr diffusion of 301 stainless steel facilitated by SMAT. Advanced Engineering Materials. 2019;**21**:1900125

[6] Abioye TE, Omotehinse IS, Oladele IO, Olugbade TO, Ogedengbe TI. Effects of post-weld heat treatments on the microstructure, mechanical and corrosion properties of gas metal arc welded 304 stainless steel. World Journal of Engineering. 2020;**17**:87-96

[7] Olugbade T, Lu J. Effects of materials modification on the mechanical and corrosion properties of AISI 316 stainless steel. In: Twelfth International Conference on Fatigue Damage of Structural Materials. Hyannis, USA: Cape Cod; 2018

[8] Olugbade T, Lu J. Improving the passivity and corrosion behaviour of mechanically surface-treated 301 stainless steel. International Conference on Nanostructured Materials (NANO 2020), Australia. 2020;**117**

[9] Olugbade TO. Electrochemical characterization of the corrosion of mild steel in saline following mechanical deformation. Analytical Letters. 2021;**54**:1055-1067

[10] Olugbade TO, Lu J. Literature review on the mechanical properties of materials after surface mechanical attrition treatment (SMAT). NanoMaterials Science. 2020;**2**:3-31

[11] Olugbade T. Datasets on the corrosion behaviour of nanostructured AISI 316 stainless steel treated by SMAT. Data-in-brief. 2019;**25**:104033

[12] Olugbade TO, Lu J. Enhanced corrosion properties of nanostructured 316 stainless steel in 0.6 M NaCl solution. Journal of Bio- and Tribo-Corrosion. 2019;**5**:38

[13] Dang C, Yao Y, Olugbade TO, Li J, Wang L. Effect of multi-interfacial structure on fracture resistance of composite TiSiN/Ag/TiSiN multilayer coating. Thin Solid Films. 2018;**653**:107-112

[14] Abioye TE, Olugbade TO, Ogedengbe TI. Welding of dissimilar metals using gas metal arc and laser welding techniques: A review. Journal of Emerging Trends in Engineering and Applied Sciences. 2017;**8**:225-228

[15] Dang C, Olugbade TO, Fan S, Zhang H, Gao LL, Li J, et al. Direct quantification of mechanical responses of TiSiN/Ag multilayer coatings through uniaxial compression of micropillars. Vacuum. 2018;**156**:310-316

[16] Olugbade TO, Abioye TE, Farayibi PK, Olaiya NG, Omiyale BO, Ogedengbe TI. Electrochemical properties of MgZnCa-based thin film metallic glasses fabricated via magnetron sputtering deposition coated on a stainless steel substrate. Analytical Letters. 2021;**54**:1588-1602

[17] Zu H, Chau K, Olugbade TO, Pan L, Chow DH, Huang L, et al. Comparison of modified injection molding and conventional machining in biodegradable behavior of perforated cannulated magnesium hip stents. Journal of Materials Science and Technology. 2021;**63**:145-160

[18] Olugbade TO. Stress corrosion cracking and precipitation strengthening mechanism in TWIP steels: Progress and prospects. Corrosion Reviews. 2020;**38**: 473-488

[19] Mohammed T, Olugbade TO, Nwankwo I. Determination of the effect of oil exploration on galvanized steel in Niger Delta, Nigeria. Journal of Scientific Research and Reports. 2016;**10**:1-9

[20] Olugbade TO, Ojo OT, Omiyale BO, Olutomilola EO, Olorunfemi BJ. A review on the corrosion fatigue strength of surface-modified stainless steels. Journal of the Brazilian Society of Mechanical Sciences and Engineering. 2021;**43**:421

[21] Olugbade TO, Olutomilola EO, Olorunfemi BJ. Review of passivity and electrochemical properties of nanostructured stainless steels obtained by SMAT: trend and progress. Corrosion Reviews. 2021

[22] Olugbade TO, Omiyale BO, Ojo OT. Corrosion, corrosion fatigue and protection of magnesium alloys: mechanisms, measurements, and mitigation. Journal of Materials Engineering and Performance. 2021. DOI: 10.1007/s11665-021-06355-2

[23] Olugbade TO, Omoniyi OO, Omiyale BO. Electrochemical Properties of Heat-Treated Al Alloy A6061-T6 in 0.5 M H_2SO_4 Solution. Journal of The Institution of Engineers (India): Series D. 2022. DOI: 10.1007/s40033-021-00313-x

[24] Kuromoto N, Simao R, Soares G. Titanium oxide films produced on commercially pure titanium by anodic oxidation with different voltages. Materials Characterization. 2007;**58**: 114-121

[25] Zhang Y, Lin T. Influence of duty cycle on properties of the superhydrophobic coating on an anodized magnesium alloy fabricated by pulse electrodeposition. Colloids and Surfaces, A: Physicochemical and Engineering Aspects. 2019;**568**:43-50

[26] Zhang L, Zhang J, Chen C, Gu Y. Advances in microarc oxidation coated AZ31 Mg alloys for biomedical applications. Corrosion Science. 2015;**91**:7-28

[27] Dziaduszewska M, Shimabukuro M, Seramak T, Zielinski A, Hanawa T. Effects of micro-arc oxidation process parameters on characteristics of calcium-phosphate containing oxide layers on the selective laser melted Ti13Zr13Nb alloy. Coatings. 2020;**10**:745

[28] Mu W, Han Y. Characterization and properties of the MgF_2/ZrO_2 composite coatings on magnesium prepared by micro-arc oxidation. Surface and Coating Technology. 2008;**202**: 4278-4284

Chapter 3

The Evaluation of the Comparative Corrosion Behaviour of Conventional and Low-Nickel Austenitic Stainless Steel: Hercules™ Alloy

Duduzile Nkomo and Nomsombuluko Masia

Abstract

Austenitic stainless steels require approximately 8% Ni to maintain austenitic microstructure at room temperature for alloys such as 304 stainless steel (304SS). Ni contributes approximately 60% of the total material cost and its price fluctuates, making the cost of austenitic stainless steel unpredictable. The use of low-nickel austenitic stainless steels as a substitute has been considered in order to remedy costs associated with Ni price fluctuations. Alloying elements such as Mn and N have been considered, however they have been found to reduce corrosion resistance. A new alloy namely Hercules™ has been developed with reduced Ni content (1.8–2% Ni). This chapter presents a comparative study of the corrosion behavior of Hercules™ and 304SS in different solutions. The alloys were evaluated using cyclic polarisation technique and immersion tests. The results demonstrated that the corrosion resistance of Hercules™ is comparable to that of 304SS. This presents the alloys as potential industrial substitutes of each other.

Keywords: low-nickel austenitic stainless steel, Hercules™ alloy, pitting corrosion, electrochemistry, cyclic polarisation

1. Introduction

Ni contributes about 60% of austenitic stainless steel manufacturing material price. This means that the price of austenitic stainless steel increases with an increase of Ni. Ni price fluctuation has led to major efforts to reduce its content in austenitic steels. Ni has been replaced with readily available, cheap elements such as Mn and N. Hercules™ is a low-Ni austenitic stainless steel alloy that was developed at Mintek-South Africa in the Advanced Materials Division. The typical content of Hercules™ comprises of 2 wt.% Ni, 9 wt.% Mn and 2.5 wt.% N [1–3].

When Hercules™ was tested for mechanical properties, it was found that it had higher tensile strength than 304SS in the hot rolled and annealed condition, hence it was termed Hercules™. This indicated that it can be used for structural applications where high strength is required. Possible applications for Hercules™ were targeted at reinforcement bars/rebars, fasteners and hot rolled channels for construction purposes. Construction industry is likely to benefit from Hercules™ because it has higher strength and reduced cost than 304SS. Other industries will soon find benefit also because it could later be available in flat product such as sheet and coil [4].

A minimum yield strength required in structural applications is 400 MPa in hot rolled condition. Typical tensile properties and density for Hercules™ alloy in hot rolled condition are shown in **Table 1** [4, 5]. Fastener prototypes of Hercules™ have been manufactured and was divided into two types; the large head bolts (M16 and M24) and roof fasteners.

According to White *et al.* [6], 240,000 tonnes of fasteners are produced globally per year and South Africa contributes only about 35,000 tonnes per year. Production of corrosion resistant bolts and roof fasteners promises a viable business. It is required that new LNASSs be resistant to corrosion. The latter requirement is in response to specific environment such as swimming pool applications, which requires pitting and crevice corrosion resistance [6].

Furthermore, about 400,000 tonnes of stainless steel rebar is produced per year. South Africa has been using about 95% carbon steel rebars which are cheaper than 304SS. However, there was little success in marketing of carbon steel rebars because of poor corrosion resistance. Corrosion of rebar is detrimental in that it can cause concrete spalling which leads to infrastructure failure. An infrastructure repair is more costly than preventing failure. Hence there is a need to use stainless steel rebar in any concrete because they are more corrosion resistant than carbon steel rebar. The conventional LNASS 201 stainless steel (201SS) is hardly available in South Africa; therefore, Hercules™ could close the gap for applications that require 304SS properties at a lower price. The cost production of Hercules™ bar is around 25% less than 304SS using a similar production route [6].

The corrosion resistance was however compromised by addition of Mn and N in Hercules™. Thus, to counteract this, 0.5 wt. % Mo was added (Hercules™ B) [2]. The focus is on characterisation of the pitting behaviour of Hercules™ B (with 0.5 wt. % Mo) against Hercules™ A (without Mo addition) and 304SS.

Pitting corrosion is the local discontinuity of a passive film which results in small holes through the material. These holes are referred to as pits. Initiation of pits can be caused by mechanical imperfection such as surface damage or inclusions. The composition of stainless steel may cause formation of inclusions which become nucleation sites for pits at the inclusion-austenite matrix interface [7]. Bautista *et al.* [8] studied the morphology of pits that were formed on the surface of test coupon after polarisation in NaCl. Pits were found to nucleate preferentially at the point of strain and at geometrical irregularities that favoured formation of corrosion cells [8].

The effect of Mo content on pitting and crevice corrosion resistance of stainless steels in chloride environments has been studied by Kaneko *et al.* [9]. The pitting potential for austenitic stainless steel alloys with 2 wt. % and 5 wt. % Mo contents showed a dramatic increase compared to that of steels without Mo in chloride environment. It is understood that Mo is adsorbed at the dissolving interfaces of a corroding metal and hence inhibiting dissolution kinetics [3, 9].

UTS (MPa)	850
0.2% Proof Strength (MPa)	500
Elongation	50%
Density (g/cm^3)	7.85

Table 1.
Typical mechanical properties of Hercules™ alloy [4].

Kaneko *et al.* [9] findings are consistent with the work that was done by Pardo *et al.* [7]. He studied the effect of both Mn and Mo in the pitting corrosion resistance of 304SS and 316SS. Tests were performed by immersion in 6 wt. % $FeCl_3.H_2O$ and cyclic polarisation technique in 3.5 wt. % NaCl. The scanning electron microscopy was used to examine the morphology of pits on the surface of corroded coupons. Images of corroded samples of 304SS that were electrochemically tested in 3.5 wt.% NaCl are shown in **Figure 1** [7].

Alloys with higher Mn content had larger pits compared to ones with lower Mn content. That is, higher Mn stainless steels experience high pitting because it was thought that since Mn has high affinity for sulphur, it reacts with Mn to form inclusions (MnS), which in turn are precursors for pit nucleation. When Mo was increased to 2.10 wt.%, pit density decreased and the size of pores evidently decreased as shown in **Figure 2** [7].

Pit initiation can also be influenced by surrounding conditions such as gaseous environment, temperature and the nature of the electrolyte. Stainless steels tend to form deep pits at specific areas when exposed to environments that contain solutions with chloride, bromide or hypochlorite [10, 11].

The occurrence of the electrochemical reactions is a result of a potential change created on a conductive metal when exposed to a conductive medium. Electrochemical potential is accompanied by electron movement which leads to electron availability at the metal surface. The electron movement or potential difference can affect the rate of corrosion reactions. Overall, the energy change provides the driving force and control for the spontaneous direction for a chemical reaction. The change in energy can be understood using thermodynamics to show how conditions of the corrosion cell can be adjusted to avoid corrosion. When a metal is immersed in a conductive solution, a charged surface of an alloy forms a complex interface. The interface is formed when the polar H_2O molecules form an oriented solvent layer. The electric field formed around the solvent layer prevents easy charge transfer, thereby limiting electrochemical reactions at the surface of an alloy [10].

The positively charged ions such as Fe^{2+} at the anode are transferred to the conductive solution which acts as an electrolyte for the cell. The electrolyte consists of ions that create electrical connectivity with the metal. Oxygen and water act on the cathodic reaction and accept negatively charged ions to form hydroxyl ions $\left(OH^-\right)$. Further anodic reactions occur simultaneously with cathodic reactions during corrosion. Typical anodic reactions are shown in Eqs. (1)–(3) and the cathodic reaction in Eq. (4). These type of corrosion reactions occur when the alternative air exposure and water is present, for example in the sea wave condition [12].

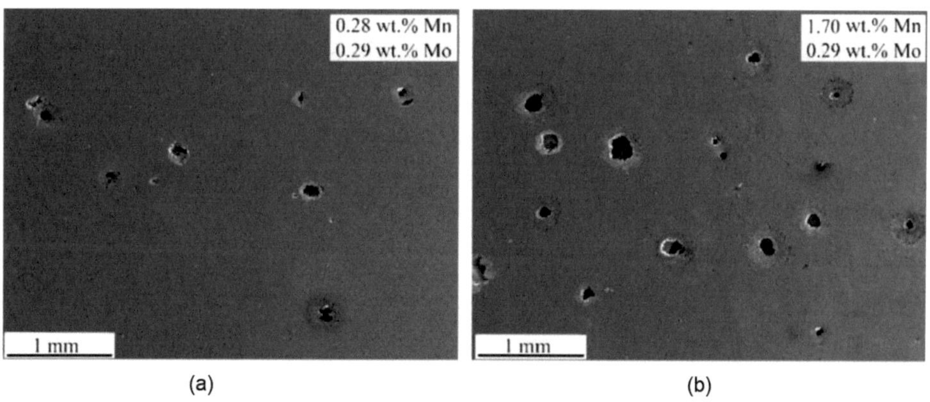

Figure 1.
Pitting corrosion of 304SS as a function of Mn (adopted from [7]).

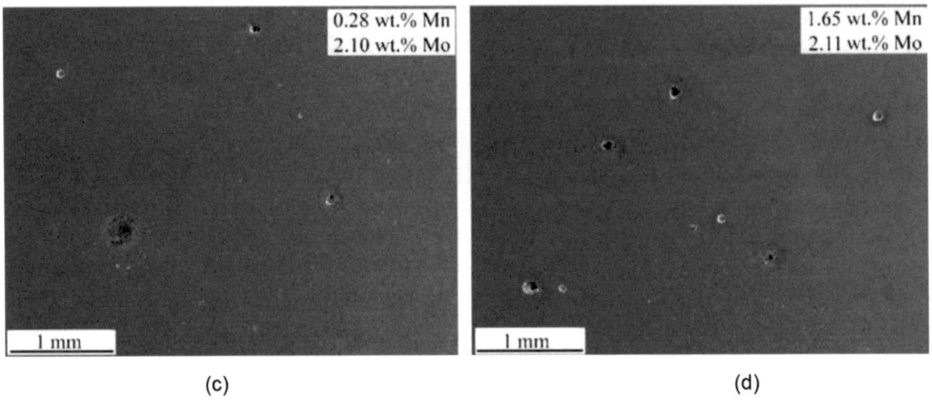

Figure 2.
Pitting corrosion of 304SS as a function of Mo (adopted from [7]).

$$Fe \rightarrow Fe^{2+} + 2e^{-} \tag{1}$$

$$Fe^{2+} + 2OH^{-} \rightarrow Fe(OH)_2 \tag{2}$$

$$4Fe(OH)_2 + 2H_2O + O_2 \rightarrow 4Fe(OH)_3 \tag{3}$$

$$4e^{-} + O_2 + 2H_2O \rightarrow 4OH^{-} \tag{4}$$

Stainless steels can corrode by pitting mechanism without a significant loss of weight on a whole structure being recognised. For example, chloride induced corrosion of stainless steel rebars in the concrete occur when there exists a difference of electric potential along the rebar. In the presence of chloride ions, the surface of the rebar is activated to act as the anode and the passivated region becomes the cathode. The reactions involved are shown in Eqs. (5) and (6) [12].

$$Fe^{2+} + 2Cl^{-} \rightarrow FeCl_2 \tag{5}$$

$$FeCl_2 + 2H_2O \rightarrow Fe(OH)_2 + 2HCl \tag{6}$$

The chloride ions migrate easily towards the interior of the pit and catalyse the hydrolysis reaction. An acidic environment is created in the pit solution as the reaction continues. During corrosion, more than one anodic reaction takes place because of different elements present in the alloy. The electrons produced by these anodic reactions are consumed by the cathodic reactions which includes hydrogen and metal reduction. Removing the cathodic sites therefore reduces the rate of corrosion. The potential required for corrosion to take place is denoted by potential corrosion E_{corr}. This is the potential at which the total rate of all anodic reactions is equal to the total rate of all cathodic reactions. The corrosion current density at this point is denoted by i_{corr} and it is used to measure corrosion rate of the metal as anionic species are released [13].

Electrochemical tests can be used to determine i_{corr} and can be measured indirectly with the aid of a counter electrode and electronic equipment. This technique uses a potentiostat in conjunction with the reference electrode. Potentiostat is an instrument that applies a potential to a specimen which enables the modification of current flow. The commonly used electrochemical techniques are polarisation resistance technique, electrochemical impedance and Tafel extrapolation. The current can be measured by extrapolation procedure whereby a specimen is initially made to act as a cathode in the electrochemical cell containing the test solution [13]. Some of these techniques can be used to determine the lifetime of a metal by calculating the time required for initiation and propagation of corrosion to cause failure.

Studies have been conducted on commercial stainless steels such as 304SS and 316SS. Pitting or crevice corrosion resistance of these alloys in chloride environments can also be measured by immersion tests in metal-chloride solutions [14].

Bergstrom *et al.* [8] followed guidelines outlined on ASTM G48 method A and B [15] to test susceptibility of 201SS and 304SS to pitting corrosion. ASTM G48 method A is the practice for measuring pitting resistance and method B for crevice corrosion resistance. These two methods included immersion of coupons in 6 wt.% $FeCl_3.6H_2O$ and measuring mass loss due to either pitting or crevice corrosion after 72 hours as shown in **Table 2**. The results showed that there was no significant difference in the mass loss of 201SS and 304SS. Therefore, this means that according to Bergstrom *et al.* [8], there was no difference in the corrosion behaviour of 201SS and 304SS in 6 wt.% $FeCl_3.6H_2O$ even in the presence of an artificial crevice.

Similarly, Garcia-Alonso *et al.* [16] also performed corrosion tests of 304SS and 316SS rebars embedded in concrete with different chloride concentrations. Tests were also done for carbon steel and LNASSs for comparison. LNASSs had Ni composition of 0.2 wt. % and 1.5 wt. % with addition of Mn. The concrete was manufactured with additions of 2 wt. % and 4 wt. % chloride content of cement. Other rebars were also embedded in a portion of concrete without chlorides and immersed into 3.5 wt.% NaCl solution to evaluate the effect of diffusion of chloride ions through the non-chlorinated concrete in the corrosion behaviour of stainless steel rebars [16].

The i_{corr} values of carbon steel, 304SS, 316SS and LNASSs were measured using electrochemical methods. In the absence of chlorides, i_{corr} values for all test alloys were measured around 0.1 $\mu A/cm^2$. The i_{corr} for carbon steel was observed to moderately increase after 30 days of immersion, which is attributed to diffusion of chlorine at the surface of the rebar. The i_{corr} values for alloys that were embedded in concrete with 2 wt. % chlorides were measured to be 3–5 times higher for carbon steel compared to that of LNASSs and 316SS. In the slab with 4 wt. % chlorides, carbon steel was measured to have i_{corr} value 10 times higher than of stainless steels. 304SS was measured to have an i_{corr} value that is of at least one magnitude lower than other alloys in 4 wt.% chlorides concrete slab [16].

The response of these stainless steel and carbon steel alloys towards increased chlorides concentration can be attributed to local breakdown of the passive layer depicting the occurrence of pitting corrosion [1, 16].

Furthermore, Bautista *et al.* [14] also performed corrosion experiments for LNASS Type 204Cu stainless steel (204SS) in a solution simulating "pore solution" of the concrete. The composition of 204SS consisted of 1.89 wt. % Ni and 8.25 wt. % Mn.

		201SS	304SS
ASTM-G48 A	Mass loss	0.0228 g/cm^2	0.0280 g/cm^2
	Max. Pit depth	0.0762 mm	0.0762 mm
ASTM-G48 B	Mass loss	0.0211 g/cm^2	0.0205 g/cm^2

Table 2.
Results of ASTM-G48 A and B tests conducted at 22°C by Bergstrom et al. [8].

Cyclic polarisation technique was used to test the susceptibility of 204SS rebars towards pitting corrosion against conventional 304SS and 316SS.

A number of mixtures of saturated calcium hydroxide concrete solutions were used with different NaCl additions. The pitting potential values were measured from the cyclic polarisation curves at the potentials where the current sharply increases when the working electrode is anodically polarised. No pitting was detected on the media without NaCl addition. However, pitting was detected for tests done with additions of NaCl. The greater the amount of NaCl added for each test, the lower the pitting potential obtained. The presence of chloride ions causes the passive layer to break down at potentials below the transpassive region and results in pitting corrosion. An example (Adapted from [14]) of pitting scans showing the effect of addition of NaCl in the corrosion behaviour of 204SS is shown in **Figure 3** [14].

The cyclic polarisation curve labelled 0% NaCl illustrates the typical behaviour of stainless steels in the absence of chloride ions. The curve shows passivity until it reaches a potential above 650 mV and evolution of oxygen is observed and referred to as transpassive region. The reverse scan for 0% NaCl curve shows current density values that are less than that of the forward scan. This means that in this medium, 204SS is not susceptible to pitting corrosion. Increasing NaCl concentration reduced the pitting potential value with a certain significant order and also increases the current density. This therefore means that the presence of chloride ions speeds up the initiation and propagation of pits. Moreover, it was noted that no repassivation or protection potential (E_{pro}) was determined for tests with NaCl additions. Comparison of corrosion behaviour of 204SS with 316SS and 304SS in the solution with 0.5 wt.% NaCl is shown in **Figure 4** [14].

The pitting potential for 204SS was measured closer to that of 304SS and 316SS around 700 mV vs. SCE. Non-carbonated solution on its own is alkaline with the pH of 12.6 and without NaCl alloys did not experience pitting corrosion. When NaCl was increased to 5 wt. %, it was observed that 204SS had significantly lower pitting potential. The difference in E_{pit} obtained at different NaCl concentrations is shown in **Figure 5** [14].

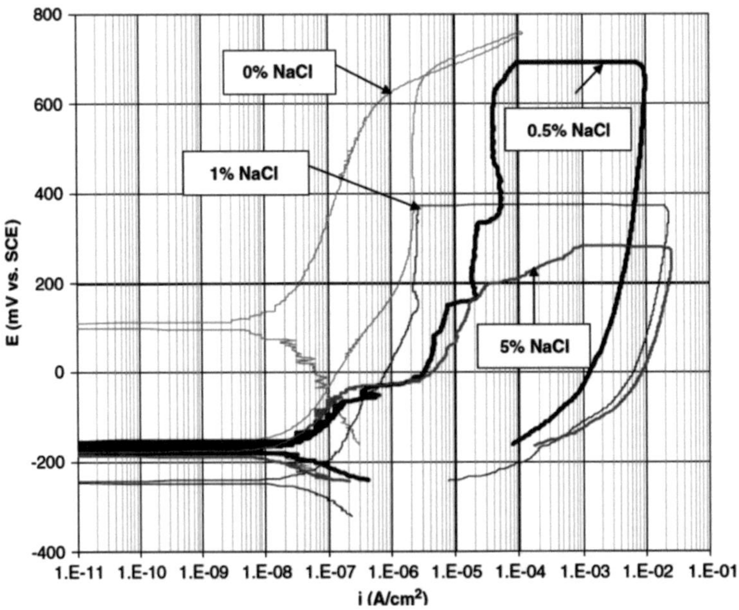

Figure 3.
Influence of chloride concentration on the pitting of 204SS (adopted from [14]).

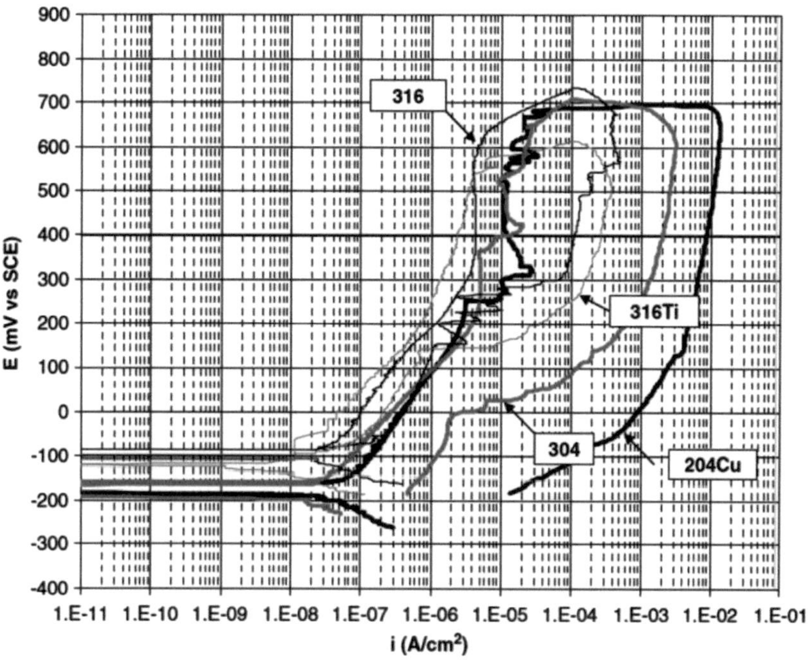

Figure 4.
Pitting of alloys in concrete with 0.5 wt.% NaCl (adopted from [14]).

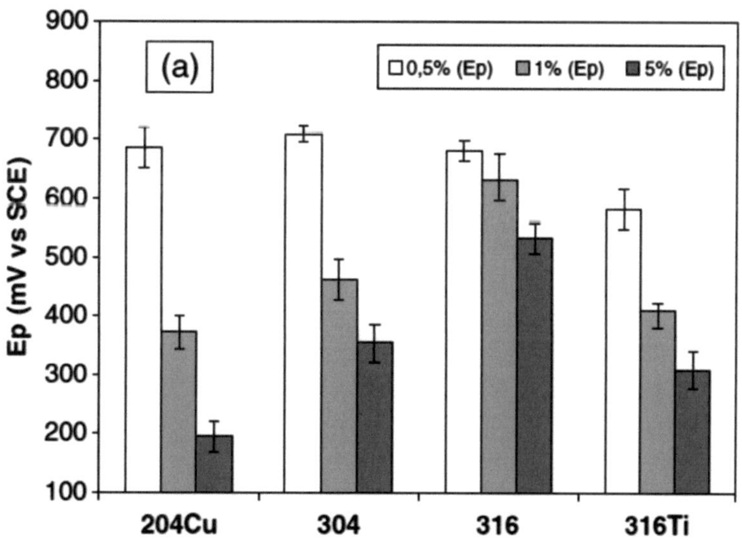

Figure 5.
Variation of pitting potential of stainless steels with variation of NaCl [14].

The general interpretation of pitting scans is: If the E_{pit} value is more noble than E_{corr} of a tested alloy, then pitting will not occur. Thus if the test solution constituents raises the pitting potential above what is estimated by the polarisation curve, that solution will protect the steel from pitting corrosion. Moreover, if the potential of the test solution is below the measured value of E_{pro}, pitting corrosion will not take place. [$Ca(OH)_2$] with 0.5% NaCl as an additive proved to be the safer one to use for concrete solution for all alloys that were tested because higher E_{pit} values were obtained compared to solutions with 1 wt.% and 5 wt.% NaCl additive [14].

Bautista *et al.* [14] studies were in agreement with work done by Berke *et al.* [17], where carbon steel rebars were tested in solutions containing different chloride concentrations. It was observed that pit nucleation became more active with increasing chloride content [17]. It is therefore important to measure the chloride content of the test solution in order to determine if the test alloy will experience corrosion in a certain chloride environment or not.

In order to determine whether certain steel will corrode in a chloride environment, a critical chloride threshold level (CCTL) is calculated [18]. The CCTL is the measure of the chloride level that is enough to cause pitting. According to Bautista *et al.* [14], the CCTL for 204SS was measured to be 1 wt.% chloride and for 304SS was 5 wt.% [14].

Stainless steel producers have also ran in-solution tests to determine the CCTL of stainless steels and carbon steel in the concrete solution. The CCTL for carbon steel was measured to be less than 0.35 wt. % chlorides and 2.51 wt. % chlorides for 304SS. Garcia-Alonso *et al.* measured the CCTL of 304SS to be 2 wt.% chlorides in similar test conditions [16, 19].

Fajardo *et al.* [20, 21] also tested LNASS (4.32 wt.% Ni) against 304SS in a carbonated concrete solution with different chloride concentrations. The cyclic polarisation curves that were obtained showed that LNASS had almost similar pitting behaviour to that of 304SS, with 304SS having a slightly higher pitting potential at all chloride concentrations.

The results that were obtained by Fajardo *et al.* [20] were in agreement with the results that have been obtained by other researchers [14–17]. LNASS was expected to have pitting potentials significantly lower than those of 304SS, but further analysis of corroded samples showed that both LNASSs and 304SS had similar behaviour.

Thus, for the current work, we study the general corrosion behaviour of Hercules™ -a LNASS alloy using standard testing methods in comparison to 304SS. Cyclic polarisation technique and immersion tests are used. The objective was to evaluate whether or not the newly developed alloy has corrosion behaviour comparable to that of 304SS and therefore can be a candidate for applications such as reinforcement bars, fasteners and hot rolled stainless steel sheets.

2. Test materials

The chemical composition of test alloys was analysed using spark emission spectroscopy. The typical composition (by weight) of both Hercules™ alloys is:

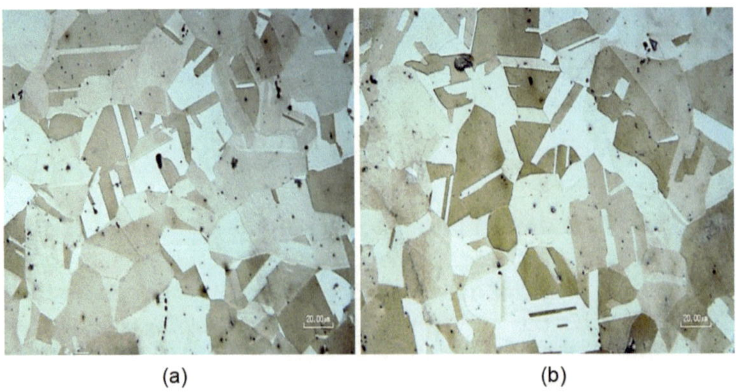

Figure 6.
The general microstructure (scale bar 20 μm) of Hercules™ a (left) and Hercules™ b (right).

2% Ni, 16% Cr, and 10% Mn and 0.25% N, with a difference on Mo content where Hercules™ A (no Mo addition) and Hercules™ B (0.5% Mo). The test samples were in the hot rolled and annealed condition, while 304SS (8%Ni, 18%Cr, 1.8%Mn, 0.03%N) was commercially received.

Figure 6 shows the general microstructures of Hercules™ alloys. Grinding, polishing and etching of the sample surface was performed in order to reveal the microstructural features. The Beraha tint etchant was used to colour and highlight the twin boundaries white and the prior austenitic grains brown. The microstructural evaluation showed that Hercules™ alloys have fully austenitic microstructure with an insignificant amount of ferrite (etched dark spots) dispersed on the austenite matrix at room temperature.

3. Corrosion tests

Table 3 represent the summary of corrosion tests that were conducted.

3.1 Cyclic polarisation

Cyclic polarisation technique was used to investigate the susceptibility of Hercules™ and 304SS alloys to pitting. This technique is generally used to measure the pitting tendencies of alloys in a given metal-solution system. The experiment starts by applying the potential scan beginning at E_{corr} and continuing in the anodic direction until there is a large increase in current (e.g. a sustained anodic current density $\geq$ 10 μA/cm^{-2}). The resulting graph is a plot of applied potential vs. the logarithm of current density. When the scan reaches the programmed current density limit value, it reverses and begins scanning in the negative direction. An example of a typical cyclic polarisation plot is shown in **Figure** 7 [22, 23].

The E_{pit} is the potential at which stable pits initiate and propagate as applied potential increases. E_{pro} is the potential below which no initiation of pits will occur. Pits that form above E_{pit} will eventually repassivate below E_{pro} hence the potential is also referred to as the repassivation potential. Both E_{pit} and E_{pro} are used to explain the kinetics of pitting and repassivation [23].

The size of the hysteresis loop can give a rough indication of the extent of propagation of initiated pits. The longer time it takes for pits to repassivate, the bigger the hysteresis loop. This imply that formed pits are severe and stable. In some cases, pits show no tendency to repassivate by the hysteresis loop closing at a potential less than E_{corr} [23].

In our work cyclic polarisation technique has been used to evaluate the corrosion behaviour of Hercules™ alloys and conventional 304SS. The procedure outlined in the ASTM G61 standard [24] was used to conduct pitting corrosion tests. The 12 mm diameter disc shaped samples were prepared from each alloy. Three test solutions were used, 3.56 wt. % NaCl, reduced concentration to 1 wt. % NaCl and 5 wt. % H_2SO_4.

Technique	Standard testing procedure	Solution
Cyclic polarisation	ASTM G61	3.56 wt.% NaCl and 1 wt.% NaCl
Cyclic polarisation	ASTM G61	5 wt.% H_2SO_4
Immersion	ASTM G31	5 wt.% H_2SO_4
Immersion	ASTM G48 A	6 wt.% $FeCl_3.6H_2O$

Table 3.
The summary of corrosion tests.

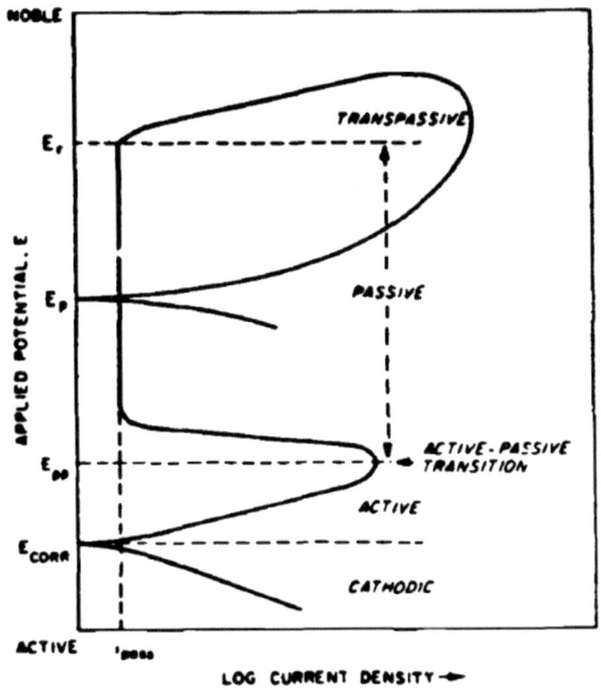

Figure 7.
A typical cyclic polarisation plot [22].

Test samples were ground to 600 grit SiC paper finish. The corrosion test conditions were set as shown in **Table 4**. An ACM potentiostat was used to apply the potential on the working electrode and to measure the current flow between the counter electrode and the working electrode. A graphite rod was used as a counter electrode and a saturated calomel electrode (SCE) was used as a reference electrode. The scan was set to increase the potential stepwise starting from the corrosion potential to 1200 mV. Duplicate scans were performed.

The corrosion potential (E_{corr}), pitting potential (E_{pit}) and protection potential (E_{pro}) were measured from the cyclic polarisation curves and they were analysed using the Origin program. The corroded coupons were further taken for analysis under stereomicroscope at 50X magnification. Information about the extent of passivation region, stability of passive state and the ability of tested alloys to spontaneously passivate in a given environmental system was obtained. The corrosion rate (*CR*) was calculated using Eq. (7) [25]. The critical current density (i_{corr}) was obtained by extrapolating the Tafel region of the cathodic and the anodic regions of the polarisation curve.

$$CR = 0.011 \times i_{corr} \times 1000 \left(mm / y \right) \tag{7}$$

Parameter	Conditions
Temperature	Room temperature
Scan rate	10 mV/min
Reverse current density	5 mA/cm^2

Table 4.
Cyclic polarisation test parameters.

3.2 Immersion tests

Immersion tests were performed in 5 wt. % H_2SO_4 for 10 days. Tests were done following the guidelines outlined in ASTM G31 [26]. 25 by 50 mm^2 coupons of Hercules™ A, Hercules™ B and 304SS were ground to 600 grit SiC paper finish. Test coupons were left for a minimum of 24 hours to allow them to passivate in air, in order to spontaneously form the passive layer possibly disrupted by sample preparation [27]. Test coupons were then immersed in a test solution at room temperature. The amount of solution in the beaker was calculated to a ratio: 0.20 ml/mm^2 [26].

ASTM G48-Method A was used for immersion tests in $FeCl_3.6H_2O$. The temperature was maintained at 26°C ± 2°C. The 25 by 50 mm^2 coupons were also prepared. Test coupons were immersed in 6 wt. % $FeCl_3.6H_2O$ for 72 hours. Then coupons were removed and rinsed with water and ethanol. Corroded coupons were then reweighed and examined for pitting [15]. Pit depth and density were measured by visual examination of images taken at 10X magnification using the Light microscope.

Mass loss due to corrosion was measured and corrosion rate calculated using Eq. (8) [26].

$$\text{Corrosion rate} = \frac{K \times W}{A \times T \times D} \tag{8}$$

Where:

$K = 8.76 \times 10^4$ for corrosion rate in millimetres per year (mm/y), W = mass loss in grams, A = area in cm^2, T = time of exposure in hours, D = density in g/cm^3.

4. Corrosion performance

4.1 Localised corrosion mechanisms in sodium chloride

Polarisation scans are shown in **Figure 8**. All test alloys performed poorly in 3.56 wt. % NaCl, with just a slightly higher E_{pit} around 72 ± 19.4 mV. SCE for 304SS. However similar corrosion behaviour was observed for all test alloys. The E_{pit} values for Hercules™ A and Hercules™ B were measured to be almost similar at more electronegative potentials. E_{pro} was measured at potentials below the pitting potential E_{pit} for all test alloys, which is an indication of severe pitting corrosion as observed by higher anodic currents at the E_{pit}, that is, there was a sudden increase of current at lower potentials. The Tafel extrapolation showed that E_{corr} values were almost similar for all test alloys and presumptuously this could be expected because of similar corrosion constants *K*.

In 1 wt. % NaCl the E_{pit} was measured at 219 ± 17.5 mV. SCE for Hercules™ A which is more positive than the one obtained from 3.56 wt. % NaCl test, as shown in **Figure 9**. The E_{pit} for Hercules™ B was measured at 579 ± 0.8 mV. SCE. There was a significant difference in E_{pit} values for each alloy due to reduction of NaCl concentration. All test alloys scans showed incapability to repassivate as the E_{pro} was measured below E_{corr}.

The behaviour of 304SS and Hercules™ in 1 wt. % NaCl can be explained by considering the effects of Cr since LNASSs have lower Cr content than 304SS. Ujiro *et al*. [28] has studied the effect of Cr additions in the corrosion behaviour of austenitic SSs in NaCl solution. It was noted that increasing Cr reduced the rate of increase of current density above E_{pit}, that is, the size of hysteresis loop was reduced. In the alloy with less Cr content, the hysteresis loop will show higher anodic currents than one with higher Cr contents as observed with 304SS and LNASSs comparisons. This demonstrates the inhibitory effect of Cr on initiation of localised corrosion.

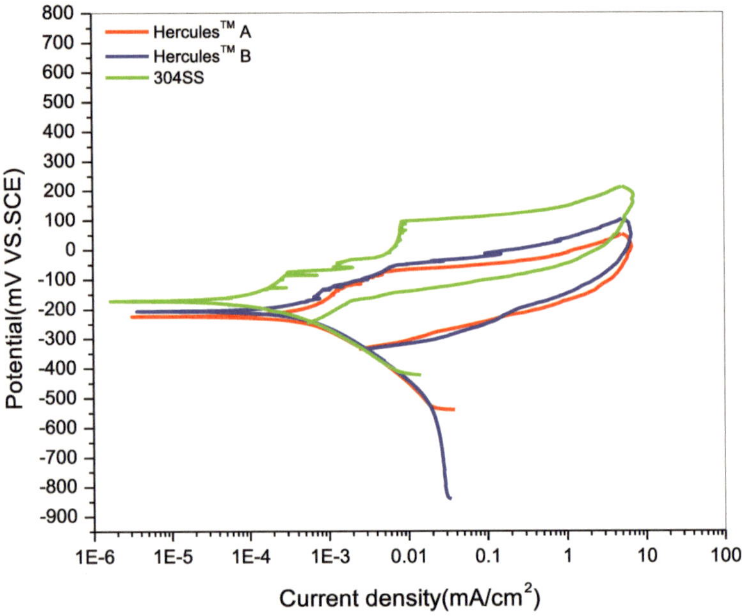

Figure 8.
Polarisation scans of austenitic SSs in 3.56 wt. % NaCl.

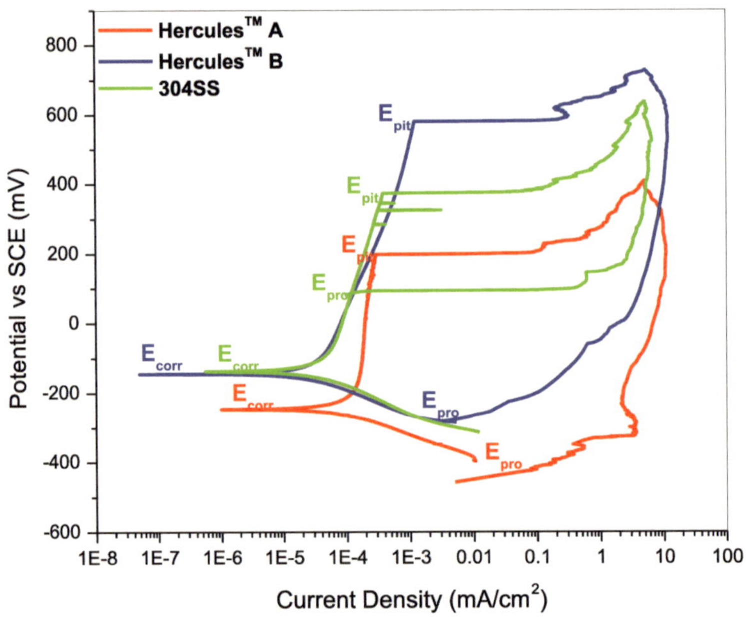

Figure 9.
Polarisation scans of austenitic SSs in 1 wt. % NaCl solutions.

Typically, in stainless steels a Cr-rich passive oxide film will form when exposed to an oxidising environment. It has been previously reported that the film consists of layers of different compositions formed due to dissimilar conditions at the metal-oxide interface and oxide-environment interface of the passive film. Depending on the composition of an alloy, it has been observed that the metal adjacent to the passive film is enriched with Ni, while the passive film itself consists of Cr-rich oxide inner layer and an Fe-rich hydroxide outer layer. The Ni-rich layer is formed because of diffusion of Fe and Cr out the bulk metal into the oxide layer. **Figure 10** shows that Cr and Mo are more significant in the formation of oxide layer because of their high affinity for oxygen. On the other hand, Ni does not participate in the formation or stabilisation of a passive layer [29].

Moreover, Ujiro *et al.* [28] studied the effect of Mo in the corrosion behaviour of austenitic SSs containing 26 wt. % Cr and varying Mo contents from 0 to 4 wt. %. It was observed that addition of Mo decreased the anodic current density; that is, the current measured before E_{pit} or before the onset of pitting. Extended passive region was evident in Hercules™ B compared to Hercules™ A and 304SS. However, both Hercules™ alloys show similar hysteresis loop behaviour. The hysteresis loop closes at potentials lower than E_{corr}.

To further substantiate polarisation results micrographs of corroded samples surface were evaluated. The micrographs of corroded samples surface are shown in **Figure 11**. It is evident from the optical micrographs that all samples experienced crevice corrosion underneath the crevice washer and as a result E_{pro} was measured at potentials below E_{pit}.

In a crevice metal-solution system, there lies a crevice critical solution, of which a minor shift in potential gradient changes the corrosion behaviour of an alloy from passive to active. The longer it takes for an alloy to reach that crevice critical solution defines the resistance of an alloy to corrosion [30]. Amongst other factors, crevice critical solution is mainly affected by alloy composition. The cationic metal species react with water to generate acidity in the crevice region. In stainless steels, the typical chemical reaction to take place under a crevice is: $Fe^{2+} + 2H_2O \rightarrow Fe(OH)_2 + 2H^+$.

The reaction involves other alloying elements such as Cr, Mn and Mo. Mo was added in Hercules™ B with an expectation that it will inhibit the localised corrosion reactions by lowering rate of generation of acidic hydrogen and consequently lower the corrosion rate. However, the reaction rate is controlled by two different kinetic phenomena. The first is charge transfer or activation control. In this case, the reaction rate is controlled by the size of the driving force, which is either hydrogen evolution or water reduction reaction. As the driving force increases, so does the reaction rate [30].

Other mechanism controlling the rate of reaction is mass transfer through the electrolyte to the electrode surface, that is, oxygen reduction reaction. Since the reaction rate is controlled by diffusion, it cannot increase indefinitely as the driving force increases. Instead, the current reaches a maximum current density which

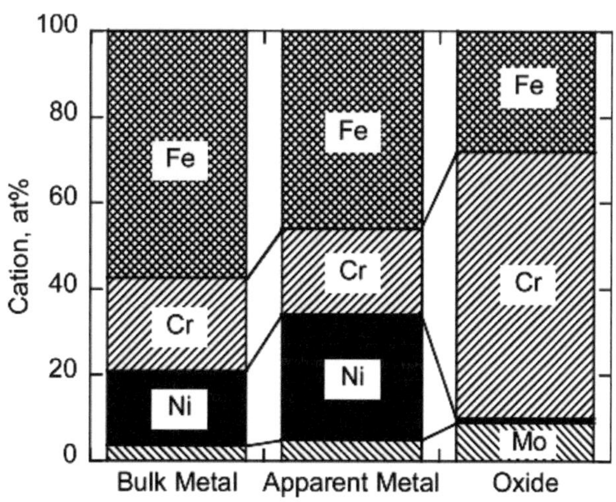

Figure 10.
Contents of optimal passive layer formed in NaCl (adopted from [29]).

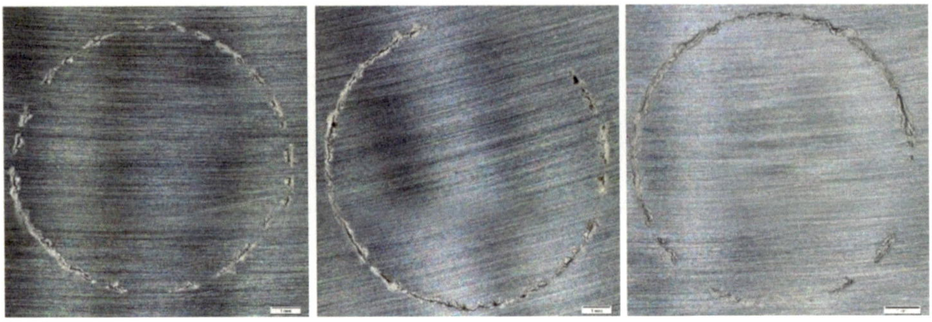

Figure 11.
Micrographs of corroded coupons (scale bar 1 mm) showing crevice corrosion in 3.56% NaCl solution for: Hercules™ a (left), Hercules™ B (middle) and 304SS (right).

is itself a function of the concentration of the species of interest in the solution as well as its diffusivity. Once the rate for a particular reaction has reached its limiting value, further increases in driving force will not result in any additional increase of the reaction rate. For this current work, addition of Mo was expected to reduce rate of propagation of corrosion by slowing down diffusion at the crevice. However, the concentration of added Mo (0.5%) was not sufficient to lower the corrosion rate in Hercules™ B and thus, similar corrosion behaviour was observed for all test alloys [29, 30]. Perhaps, higher amount of Mo should be added, but bearing in mind cost-related issues. Micrographs are in agreement with the polarisation scan measurements, which showed similar corrosion behaviour.

The difference was observed with micrographs obtained from 1 wt. % NaCl test, as shown in **Figure 12**, which showed pitting corrosion within the crevice area. Ujiro *et al.* [28] explained that crevice corrosion can occur either by depassivation or pitting. Depassivation type occurs by corrosion of surface underneath the crevice, due to pH drop and extensive destruction of the passive film under the crevice washer (observed with 3.56 wt. % NaCl). Pitting type occurs by pitting inside the crevice area as a result of chloride concentration increase in the inner solution.

Ujiro *et al.* [9] also investigated the relationship between the type of corrosion and the E_{corr}. It was observed that ferritic alloys which corroded by depassivation had lower E_{corr} than the ones that corroded by pitting. Crevice corrosion by depassivation is related to E_{corr} because it involves an intensive metal dissolution at lower potentials and not just a single pit. Similar to the current work, E_{corr} values obtained from 1 wt. % NaCl were measured to be higher than those obtained from 3.56 wt. %

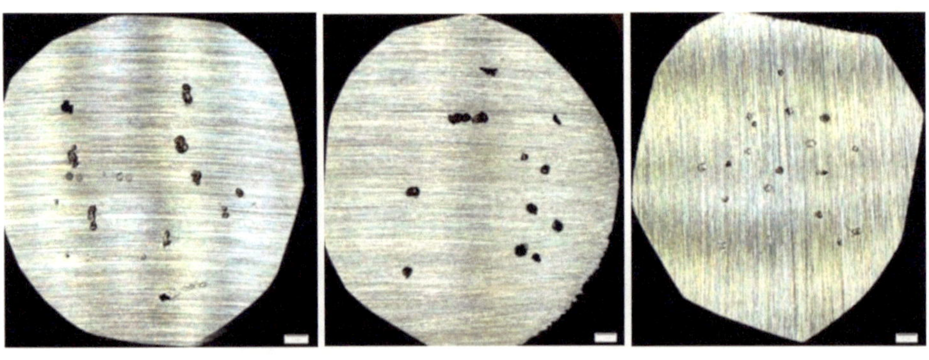

Figure 12.
Micrographs of corroded coupons (scale bar 1 mm) showing pitting corrosion in 1 wt. % NaCl solution for: Hercules™ A (left), Hercules™ B (middle) and 304SS (right).

NaCl. This means that 1 wt. % NaCl did not contain a critical chloride concentration required to reach crevice critical solution for the complete dissolution of a metal underneath the crevice. Hence, instead of depassivation, pitting was observed. The solution in the pit is more aggressive once pitting has started as shown by an increase of current density. In some alloys, once pitting initiates, propagation is faster and it becomes difficult for a formed pit to repassivate as observed in the CP scans. However, the onset of pitting was delayed for Hercules™ B in 1% wt. % NaCl, as indicated by an extended passive region than in other test alloys, owing to the inhibiting effect of Mo.

4.2 Severe pitting in ferric chloride

The mass loss of test alloys was measured and the corrosion rate due to pitting was calculated. No significant difference in mass loss of all test alloys was observed, with corrosion rates measured to be 11.8 ± 0.2 mm/y for Hercules™ A, 12.6 ± 0 mm/y for Hercules™ B and 14.1 ± 0.1 mm/y for 304SS.

This behaviour is similar to results that were obtained by Bergstrom *et al.* [8] when 201SS and 304SS were tested in 6 wt.% $FeCl_3$ for 72 hours. Both alloys showed a corrosion rate of 0.0228 g/cm^2 and similar pit depth of 0.0762 mm [8].

The 6 wt. % $FeCl_3.6H_2O$ solution is generally used as a test for localised corrosion for accelerated tests. Therefore, test alloys were immersed in the solution for 72 hours as recommended in the ASTM G84 standard. However, this test solution is used to simulate a very rough composition environment within a localised corrosion site in a stainless steel. It can be very aggressive for low-alloyed steels such as 304SS and 201SS. This has also been confirmed by tests that were conducted by Ujiro *et al.* [28]. Alloys with higher Mo and Cr contents (above 26 wt.% Cr and 4 wt. % Mo) showed more corrosion resistance than the ones with the compositions approximately similar to that of 304SS [28].

Furthermore, 6 wt. % $FeCl_3.6H_2O$ serves as a chemical potentiostat by forming the Fe^{3+}/Fe^{2+} redox couple which has an approximate potential of 450 mV with high chloride concentration. The solution is highly acidic with a pH of 1.44, which is enough to create a large current without a need to polarise the specimen as with the electrochemical tests [27]. From the electrochemical tests, it has been established that chloride concentration from neutral 1 wt. % NaCl was enough to cause a E_{pit} that is less than 450 mV. The 6 wt. % $FeCl_3.6H_2O$ solution has higher chloride concentration than NaCl, hence Hercules™ alloys and 304SS corroded aggressively. The high potential of the test solution almost guarantees that the pitting potential of each alloy was exceeded. The acidic nature of $FeCl_3.6H_2O$ also inhibits repassivation and lowers passive film strength by cathodic reactions that occur on the surface of the sample via Fe^{3+}/Fe^{2+} ions [27].

The difference in the individual pit morphology was observed. Hercules™ A had irregular shaped pits and with high depth, whilst Hercules™ B had wide and round shallow pits. This means that pitting propagated quicker in Hercules™ A than in Hercules™ B. **Figure 13** shows the one of the deepest representative pit that was observed in Hercules™ A after 72 hours of immersion in $FeCl_3.6H_2O$, along with the pits measurements. Although pit density of all test alloys was almost similar, Hercules™ A showed severe pitting because of larger pit opening.

Figure 14 shows that the size of pit opening observed in Hercules™ B is smaller than that of Hercules™ A. **Figure 15** shows the extent of pitting that was observed for 304SS. Overall, the pit evaluation proves that $FeCl_3.H_2O$ is an aggressive solution for testing LNASSs and 304SS. Even the addition of 0.5 wt. % Mo for Hercules™ B is not enough for it to be used for applications in aggressive environments.

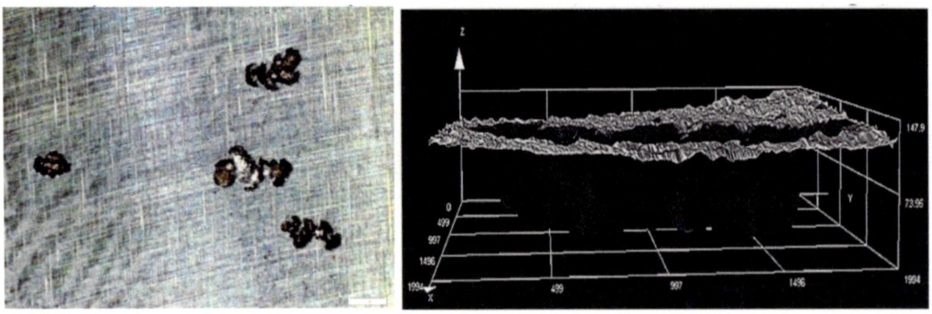

Figure 13.
Micrographs (scale bar = 1 mm) and representative pit of Hercules™ A in $FeCl_3.H_2O$.

Average	**1557 ± 206**	**291 ± 77**	**1587 ± 199**
Max.	1862	374	1881
Min.	1259	165	1289
Range	603	208	592
No.	**Width[μm]**	**Height[μm]**	**Length[μm]**
1	1424	372	1472
2	1676	165	1684
3	1563	374	1607
4	1862	266	1881
5	1259	275	1289

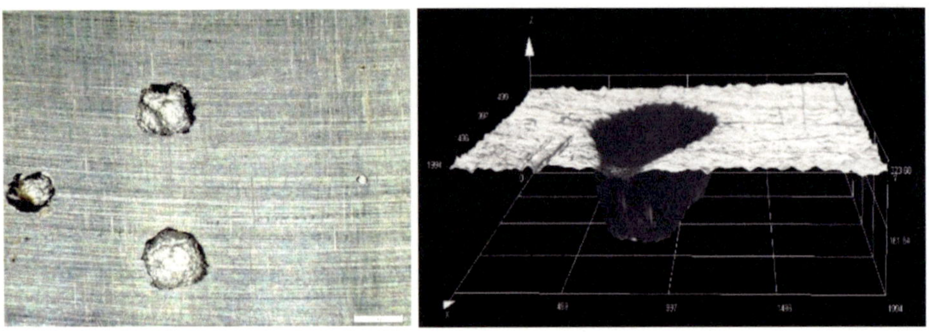

Figure 14.
Micrographs (scale bar = 1 mm) and representative pit of Hercules™ B in $FeCl_3.H_2O$.

Average	**997 ± 332**	**114 ± 7**	**1005 ± 328**
Max.	1388	129	1394
Min.	379	109	395
Range	1008	20	999
No.	**Width[μm]**	**Height[μm]**	**Length[μm]**
1	379	109	395
2	1073	110	1079
3	1073	110	1079
4	1073	110	1079
5	1388	129	1394

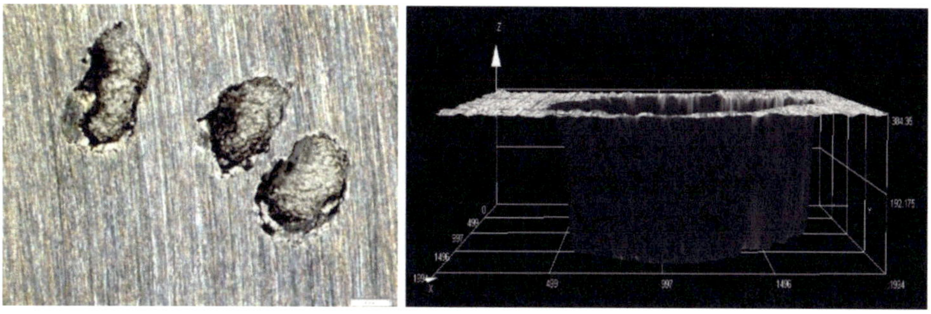

Figure 15.
Micrographs (scale bar = 1 mm) and representative pit of 304SS in $FeCl_3.H_2O$.

Average	**1552 ± 191**	**274 ± 76**	**1592 ± 150**
Max.	1863	372	1772
Min.	1324	145	1325
Range	539	227	447
No.	**Width[μm]**	**Height[μm]**	**Length[μm]**
1	1324	324	1576
2	1678	145	1683
3	1556	372	1604
4	1836	255	1772
5	1368	275	1325

4.3 Passivity behaviour in sulphuric acid

A stainless steel can be considered resistant to uniform corrosion in a particular environment if the corrosion rate does not exceed 0.1 mm/y [19] . In the current work all test alloys demonstrated resistance in 5 wt. % H_2SO_4 as shown by polarisation curves in **Figure 16**.

All alloys displayed the ability to passivate spontaneously in 5 wt. % H_2SO_4. Lower E_{corr} for Hercules™ A can be attributed to the kinetics of passivity for stainless steels, whereby if the cathodic reaction becomes more dominant at lower potentials and thus remaining in the active region will result in favourable conditions for anodic reactions to overtake the redox reactions at lower potentials. Polarisation scans of some stainless steel will show these cathodic reactions by the presence of anodic current peak, which is an indication of non-uniform passivity due to less OCP test times. Thus, it can be observed that with all test alloys, if given enough time to form a passive layer during OCP tests, the anodic peak current can be avoided and alloy are fully passivated with no disruption of the protective film when exposed to H_2SO_4. Therefore, H_2SO_4 is considered a safe environment for LNASSs and so is 304SS.

Furthermore, the absence of a hysteresis loop is an indication that tested alloys did not undergo any type of localised corrosion even though an artificial crevice was introduced in each sample. The presence of an artificial crevice creates passivation current (i_{pass}) (Current density at the passive region) that is higher than i_{corr}. However, the conditions were not sufficient to activate sample surface for formation of pits or cause crevice corrosion since the test solution did not contain chlorides [31].

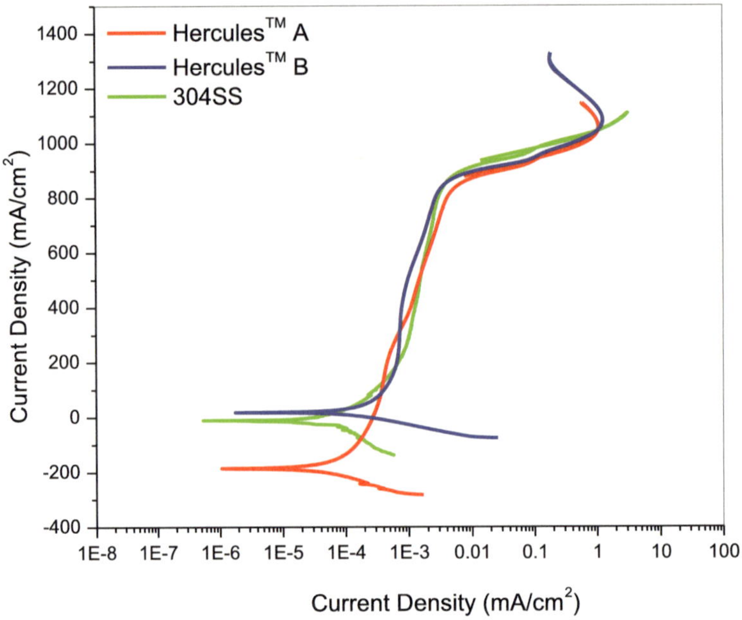

Figure 16.
Cyclic polarisation scans of austenitic SSs in H_2SO_4.

The corrosion rates calculated from polarisation curves for Hercules™ B and 304SS were comparable at 0.001 ± 0.008 mm/y and 0.002 ± 0.004 mm/y, respectively. The corrosion rate of Hercules™ A was measured to be 0.016 ± 0.029 mm/y, which is a magnitude higher than 304SS and Hercules™ B.

However, the corrosion rate of test alloys in 5 wt. % H_2SO_4 was calculated to be higher in the immersion tests. The corrosion rate of Hercules™ A was calculated from the mass loss incurred to be 1.863 ± 0.028 mm/y. Hercules™ B and 304SS had the corrosion rate less than 0.100 ± 0.020 mm/y, with 304SS having the lowest. It is often assumed that corrosion rate of stainless steels is linear with the function of time during immersion tests, but this is not always true for some stainless steels immersed for a longer time [32].

Hercules™ A was observed to react aggressively for the first 24 hours at a presumably higher corrosion rate but the vigorous reaction decreased as days progressed. Hercules™ B did not react aggressively in the beginning and throughout the entire exposure time [32].

Based on observations made in the current project, the solution in which Hercules™ A was immersed had a dark bluish precipitates after a few hours of immersion and the reaction was more aggressive than Hercules™ B. The solution with Hercules™ B and 304SS did not show any change of colour and the reaction was less aggressive. Therefore, it can be established that passive films formed by stainless steels may be broken down in a prolonged exposure period and hence higher corrosion rates are obtained in immersion tests than polarisation tests. Although mass loss is negligible after a certain period it can still add up to the final mass loss measurement. Electrochemical tests took less than 4 hours to obtain a complete scan. Therefore, the corrosion rate measured tend to be less than that obtained from immersion tests [27].

5. Conclusion

Hercules™ alloy was developed with an aim to reduce cost of austenitic stainless steel and for use as an alternative to 304SS with target applications being reinforcement bars and fasteners, taking advantage of higher strength of Hercules™. In order to fast forward acceptability for this alloy, additional applications were proposed where corrosion resistance is of importance such as rolled sheets for manufacturing of water tanks and other corrosion resistant products.

The general corrosion behaviour of Hercules™ alloy and 304SS have been evaluated using cyclic polarisation technique and immersion tests. Polarisation curves of NaCl tests showed that in higher chloride content of 3.56 wt. %, all test alloys corroded severely. When the concentration was reduced to 1 wt. %, the passive region of Hercules™ B was significantly extended, making this alloy more resistant to initiation of localised corrosion. However, once pitting initiated, it propagated faster, which was observed by the size of hysteresis loop and confirmed by visual examination of corroded coupons. This nullifies the anticipated effects of Mo and this has been explained using thermodynamics of corrosion. Immersion tests in $FeCl_3$ also proved to be extremely aggressive making these alloys not suitable for use in aggressive chloride environments such as swimming pools and sea water. Additional polarisation and immersion tests in H_2SO_4 showed that all test alloys had an ability to spontaneously passivate in this environment, thus making Hercules™ useful in such reducing environments at room temperatures.

It is however, arguable whether or not Hercules™ alloy (with 0.5% Mo) can be used as a substitute for 304SS for when Ni prices are high because the tests conducted here did not provide sufficient rigour to validate the corrosion resistance of Hercules™ against 304SS. Thus, this leaves a gap for further work to be conducted. The work presented here simply outlines the comparative behaviour of Hercules™ to 304SS and more corrosion tests via in-situ environments should be performed in order to qualify Hercules™ alloy. Temperature, composition and test parameters are other factors that can be investigated further in order to perform application-based tests. Evidently, the solutions outlined by the corrosion test standards are too aggressive for these alloys. Building from this work, a more suitable test procedure can be developed and thus, providing newly developed LNASSs a chance to demonstrate their corrosion resistant strengths, as already observed with 1 wt. % NaCl and H_2SO_4 tests.

Acknowledgements

Authors would love to thank:

University of Cape Town (Centre of Materials Engineering).

Professor R. Knutsen for the overall supervision provided throughout the entire project.

Mintek-Advanced Materials Division (Physical metallurgy group).

Ms. M. Smit for her assistance with executing electrochemical corrosion tests in the AMD corrosion laboratory.

Special thanks to Mintek for funding the project.

Author details

Duduzile Nkomo* and Nomsombuluko Masia
Mintek, Randburg, South Africa

*Address all correspondence to: dudunk@mintek.co.za

References

[1] Kerr J, Moema J, Scheers P, Paton R, Main Technical Reports on New Generation Low-Nickel (< 2%) Austenitic Stainless Steels for Structural Applications, Mintek Annual Report-C4105M, Mintek, Randburg, South Africa. 2004

[2] Moema JS, Paton R, Papo MJ. On the Development of a New Low-Nickel Stainless Steel for Structural and Long Product Applications, Mintek Publication-C4076M, Mintek, Randburg, South Africa. 2005

[3] Muwila A, Papo MJ. A more corrosion resistant Hercules ™ alloy. Southern African Institute of Mining and Metallurgy. 2007;**107**:155-157

[4] Mintek. Technical Data Sheet Stainless steel: Chromium-Nickel-Manganese Hercules(TM). Randburg, South Africa: Mintek; 2022. https://www.mintek.co.za

[5] Kerr J, Paton R. Preliminary investigations of low-nickel stainless steels for structural applications. In: Tenth International Ferroalloys Congress; 2004. pp. 757-765

[6] White RT. New generation low-nickel (<2%) austenitic stainless steels for structural applications, Mintek Commercialisation plan-Project 32434, Mintek, Randburg, South Africa. 2005

[7] Pardo A, Merino MC, Coy AE, Viejo F, Arrabal R, Matykina E. Pitting corrosion behaviour of austenitic stainless steels - combining effects of Mn and Mo additions. Corrosion Science. 2008;**50**(6):1796-1806

[8] Bergstrom DS, Botti CA. AL 201HP TM (UNS S20100) alloy: Alternative to 300 series alloys, Stainless Steel World 2005 Conference, Maastricht, The Netherlands. 2006

[9] Kaneko M, Isaacs HS. Effects of molybdenum on the pitting of ferritic- and austenitic-stainless steels in bromide and chloride solutions. Corrosion Science. 2002;**44**(8): 1825-1834

[10] Jones DA. Denny-a-Jones-Principles-and-Prevention-of-Corrosion. Environments 2nd ed, United States: University of Nevada. 1992;334:357

[11] Koumya Y, Idouhli R, Khadiri M, Abouelfida A, Aityoub A, Benyaich A and Romane A. Pitting corrosion and effect of Euphorbia echinus extract on the corrosion behavior of AISI 321 stainless steel in chlorinated acid. Corrosion Reviews, 2019;**37**(3):259-271

[12] Neville A. Chloride attack of reinforced concrete : An overview. Materials and Structures. 1995;**28**:63-70

[13] Sedriks AJ. Corrosion of Stainless Steels. Canada: John Wiley and Sons, Inc.; 1979. pp. 9-21

[14] Bautista A, Blanco G, Velasco F. Corrosion behaviour of low-nickel austenitic stainless steels reinforcements: A comparative study in simulated pore solutions. Cement and Concrete Research. 2006;**36**(10): 1922-1930

[15] ASTM G. 48-03. Pitting and crevice corrosion resistance of stainless steels, USA. 2003

[16] García-Alonso MC, Escudero ML, Miranda JM, Vega MI, Capilla F, Correia MJ, Salta M, Bennani A, González JA. Corrosion behaviour of new stainless steels reinforcing bars embedded in concrete. Cement and concrete research. 2007;**37**(10):1463-71

[17] Berke N. The Use of Anodic polarization to Determine the

effectiveness of Calcium Nitrite as an Anodic Inhibitor. Corrosion Effect of Stray Currents and the Techniques for Evaluating Corrosion of Rebars in Concrete. West Conshohocken, United States: ASTM International; 1986

[18] Ogunsanya IG, Hansson CM. Detection of the critical chloride threshold of carbon steel rebar in synthetic concrete pore solutions. RILEM Technical Letters. 2018;**3**(2018):75-83

[19] Outokumpu. Corrosion Handbook. 11th ed. Finland: Outokumpu Oyj. 2015

[20] Fajardo S, Bastidas DM, Criado M, Bastidas JM. Electrochemical study on the corrosion behaviour of a new low-nickel stainless steel in carbonated alkaline solution in the presence of chlorides. Electrochimica Acta. 2014;**129**:160-70

[21] Fajardo S, Bastidas DM, Criado M, Romero M, Bastidas JM. Corrosion behaviour of a new low-nickel stainless steel in saturated calcium hydroxide solution. Construction and Building Materials. 2011;**25**(11):4190-4196

[22] Siebert OW. Laboratory Electrochemical Test Methods. Laboratory Corrosion Tests and Standards. ASTM STP 866. MO, United States: St. Louis, Creve Coeur; 1985. pp. 65-90

[23] Princeton Applied Research, Electrochemistry and Corrosion: Overview and Technique, Princeton Applied Research. Oak Ridge: S. Illinois Avenue. 2022. www.princetonappliedresearch.com

[24] ASTM G61-86 (Reapproved 2014). Standard Test Method for Conducting Cyclic Potentiodynamic Polarization Measurements for Localised Corrosion Susceptibility of Iron-, Nickel-, or Cobalt-Based Alloys. West Conshohocken, United States: ASTM International; 2014. pp. 1-9

[25] Standard AS. G102-89, Standard Practice for Calculation of Corrosion Rates and Related Information from Electrochemical Measurements. Annual Book of ASTM Standards. West Conshohocken, PA: ASTM International. 2006:3

[26] NACE A. ASTMG31-12a, Standard Guide for Laboratory Immersion Corrosion Testing of Metals. ASTM International. 2012;100:19428-2959

[27] Baoian JSR, Dean Jr. S, Hack H. Corrosion Tests and Standards. Application and Interpretation, 2nd edn. United States: ASTM International, West Coshohocken; 2010

[28] Ujiro T, Yoshioka K, Staehle RW. Differences in corrosion behavior of ferritic and austenitic stainless steels. Corrosion. 1994;**50**(12):953-962

[29] Olsson COA, Landolt D. Passive films on stainless steels - chemistry, structure and growth. Electrochimica Acta. 2003;**48**(9 SPEC):1093-1104

[30] McCafferty E. Introduction to corrosion science. New York, USA: Springer Science & Business Media; 2010

[31] Jegdić B, Dražić DM, Popić JP. Open circuit potentials of metallic chromium and austenitic 304 stainless steel in aqueous sulphuric acid solution and the influence of chloride ions on them. Corrosion Science. 2008;**50**(5): 1235-1244

[32] Muwila A. The Effect of Manganese, Nitrogen and Molybdenum on the Corrosion Resistance of a Low Nickel (< 2 Wt %) Austenitic Stainless Steel. Johannesburg, South Africa: University of Witwatersrand; 2006

Chapter 4

Dependence of Corrosion Resistance of Austenitic Chromium-Nickel Steels on the Magnetic State of Austenite

Gennadii Snizhnoi

Abstract

Corrosive behavior of austenitic chromium-nickel steels from the magnetic state (parameter χ_0) of austenite, pre-formed to interact with aggressive media are research. Correlation between the rate K of pitting corrosion and the specific magnetic susceptibility χ_0 of austenite was experimentally established. It is experimentally established that the corrosion resistance of austenitic steels AISI304, 08Cr18Ni10, AISI 321, 08Cr18Ni10Ti (containing a low amount of δ-ferrite ~0.005 ... 0.5%) depends on the magnetic state of austenite: the corrosion rate of steel decreases with increases χ_0 austenite. The tendency of change in the corrosion rate of austenitic alloy with a high nickel content 06Crh28NiMoCuTi (not contain δ-ferrite) has the opposite character: with increasing χ_0, the corrosion rate of the alloy increases is revealed. For austenitic chromium-nickel steels, the corrosion rates of the individual (austenite (A), δ-ferrite (F), strain-induced α′-martensite (M)) and total ($A + F$, $A + M$ and $A + F + M$) phases are determined. It is proposed to predict corrosion according to the specific magnetic susceptibility χ_0 of austenite and the amount δ-ferrite.

Keywords: austenite, steel, specific magnetic susceptibility, ferrite, martensite, plastic deformation, corrosion

1. Introduction

Progress in increasing competitiveness, i.e. the application of operational and more cost-effective control of corrosion properties, requires new approaches to the study of austenitic steels. In general, these steels are studied in terms of the atomic-crystalline state of austenite, as the main component of their structural state, namely, the parameters of the crystal lattice, grain size, the presence of packaging defects and dislocations, and so on. However, a comprehensive assessment of properties for one parameter is complicated, and in some cases impossible.

Since the bulk of these steels is paramagnetic austenite, and with a certain chemical composition or hardening, its amount can reach 100%, it is proposed to study these steels from the standpoint of the atomic-magnetic state. That is, to consider austenite primarily as a paramagnetic, which is characterized by a peculiar electronic structure, for which the supersensitive parameter is the specific

IntechOpen

paramagnetic susceptibility χ_0 (the total magnetic moment per unit mass of austenite at a single value of the magnetic field). In this case, the parameter χ_0 is an integral characteristic of the formed austenite due to various factors (chemical composition, smelting conditions, deformation, heat treatment, etc.). It should be noted that most studies are devoted to the study of the magnetic susceptibility of χ steel (not χ_0 austenite), which may contain both austenite and α-phase: δ-ferrite, α′-martensite of deformation (after cold compressive plastic deformation).

There is almost no information about the relationship between the magnetic state of austenite and corrosion properties. Thus, the assumption of predicting the corrosion properties of austenitic steels by the structural-magnetic state (parameter χ_0) of austenite, which does not contain or contains a very low amount of α-phase (δ-ferrite, α′-martensite), requires theoretical, experimental and practical confirmation.

Establishing a correlation between the atomic and magnetic state of austenite and corrosion is of theoretical and practical importance. Some of the information related to this problem is debaTable and insufficiently researched.

It is known from scientific sources that the low content (~0.005 ... 0.5%) of the α-phase can dramatically affect the corrosion properties, which necessitates the introduction of supersensitive methods for its determination. The question of the influence of individual phases, which are simultaneously present in the austenitic matrix, remains unresolved.

The corrosion properties of chromium-nickel steels are affected by the casting temperature, chemical composition, carbides, heat treatment, deformation, harmful impurities, etc. No single physical parameter has been identified that would simultaneously and unambiguously take into account the influence of these factors on the properties of these steels. Therefore, for this purpose, a sensitive parameter is proposed—the specific paramagnetic susceptibility of χ_0 austenite.

Therefore, when determining the performance of austenitic steel products, along with the study of the crystal structure, it is advisable to study the relationship between the atomic and magnetic state of austenite with the corrosion properties.

2. Relationship between corrosion resistance of austenitic chromium-nickel steels AISI 304, 08Cr18Ni10, AISI 321, 08Cr18Ni10Ti (contain low amount of δ-ferrite) and paramagnetic state of austenite

Despite numerous studies of corrosion processes of various materials [1–4], the problems that arise require a comprehensive study of the mechanisms of corrosion behavior to ensure reliable and safe operation of modern structural materials. It should be noted that even in one smelter of industrial steel δ-ferrite is distributed unevenly, for example for steel 08Cr18Ni10Ti—the width of the sheet perpendicular to the rolling the amount of δ-ferrite varied from 0.01% to 0.4%. Chromium-nickel austenitic steels contain little δ-ferrite, which due to the small contact surface cannot significantly affect the corrosion behavior of austenitic steels [5]. We can assume that corrosion is affected by the pre-formed atomic-magnetic state of the whole austenitic matrix (austenite), which is characterized by the sum of magnetic moments of atoms per unit mass (specific magnetization) and described by an external parameter—specific magnetic susceptibility χ_0. In [6] it was noted that the presence of a low content of δ-phase ($P_\delta < 2\%$) in steel 12X18H10T is not detected by radiography, metallographic methods show its presence, and the phase content was determined by the magnitude of the magnetization.

To study the dependence of corrosion resistance on the paramagnetic state of the austenitic matrix (parameter χ_0) and evaluate the effect of low δ-ferrite content on corrosion behavior, two groups of the most common steels were selected: AISI 304, AISI 321 and their analogues 08Cr18Ni10, 08Cr18Ni10Ti in **Table 1**.

The samples were made of rolled sheet metal with a thickness of 1 mm of industrial melts of these steels. Taking into account that the uneven distribution of δ-ferrite over the width of the cold-rolled sheet [5] and for further averaging of the obtained results 7… 10 flat samples with dimensions of $\sim 7 \times 3 \times 1$ mm^3 were cut from different places of the sheet of each smelting. Specific magnetic susceptibility χ was determined by a magnetometric unit (Faraday balance). The content of δ-ferrite and the specific magnetic susceptibility χ_0 of austenite (according to [7]) of these samples were determined. The average content of δ-ferrite P_δ and χ_0 for each melt was calculated. The obtained results (P_δ and χ_0) [8] and corrosion rate K (attracted from the source [9]) in chloride-containing solution are given in **Table 2**.

Analysis of experimental hyperbolic dependences of the corrosion rate K [9] on the specific magnetic susceptibility of χ_0 austenite steels AISI 304, 08Cr18Ni10 and AISI321, 08Cr18Ni10Ti shows (**Figure 1**) [8]: that the larger the value χ_0, the higher the corrosion resistance (lower corrosion rate K). Since χ_0 determines the atomic-magnetic structure of austenite, it can be assumed that the corrosion behavior of austenitic chromium-nickel steels depends on the atomic-magnetic (paramagnetic) state of austenite, pre-formed to interact with the chloride-containing medium.

Steel	Melt	C	N	Mn	Si	S	P	Cr	Ni	Ti
AISI 304	1	0.071	0.048	1.23	0.22	0.001	0.027	17.96	9.34	—
		Σ0.119								
	2	0.067	0.046	1.74	0.50	0.001	0.028	18.22	8.09	—
		Σ0.113								
	3	0.075	0.055	1.65	0.43	0.004	0.024	18.25	8.09	—
		Σ0.130								
	4	0.050	0.044	1.70	0.41	0.002	0.028	18.30	8.10	—
		Σ0.094								
	5	0.030	0.039	1.81	0.39	0.001	0.034	18.10	8.20	—
		Σ0.069								
08Cr18Ni10	6	0.060	—	1.34	0.32	0.006	0.035	17.44	9.77	0.03
AISI 321	1	0.035	0.012	1.66	0.54	0.001	0.026	17.10	9.10	0.32
		Σ0.047								
	2	0.060	0.011	1.59	0.66	0.002	0.027	16.43	9.14	0.34
		Σ0.071								
	3	0.064	0.012	1.22	0.52	0.001	0.026	17.43	9.70	0.41
		Σ0.076								
	4	0.030	0.013	1.62	0.41	0.002	0.028	17.41	9.24	0.31
		Σ0.043								
	5	0.040	0.013	1.70	0.49	0.001	0.026	17.70	9.10	0.35
		Σ0.053								
08Cr18Ni10Ti	6	0.070	—	1.70	0.49	0.007	0.027	17.97	10.46	0.46

Table 1.
Chemical composition of the investigated steels, % wt.

Steel	Melting number	P_δ, %	χ_o, 10^{-8} m^3/kg	K, $g/(m^2 h)$
AISI 304	1	0.0335	2.27	47.65
	2	0.0448	2.25	57.76
	3	0.0138	2.23	86.64
	4	0.1130	2.31	43.37
	5	0.0329	2.24	67.87
08Cr18Ni10	6	0.0054	2.23	98.19
AISI 321	1	0.0453	2.55	113.0
	2	0.160	2.56	98.6
	3	0.2920	2.59	89.9
	4	0.1760	2.68	76.8
	5	0.036	2.54	126.1
08Cr18Ni10Ti	6	0.0142	2.52	165.2

Table 2.
The average values of P_δ, χ_o, K *of each investigated steelmaking.*

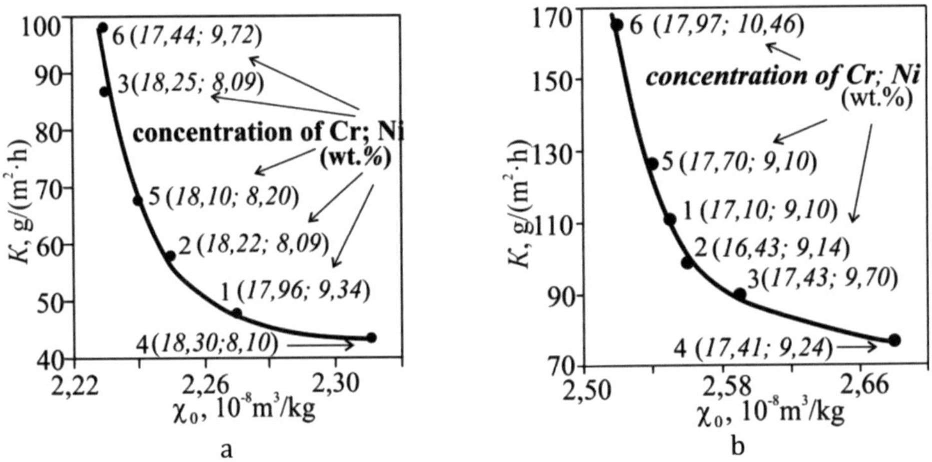

Figure 1.
Relationship between the corrosion rate K *and the specific amagnetic susceptibility* χ_o *of austenite of the studied steels. a—AISI 304 (melt 1–5) and 08Cr18Ni10 (melt 6); b—AISI 321(melt 1–5) and 08Cr18Ni10Ti (melt 6).*

In [4], an electron-probe analysis of the distribution of chemical elements for two local regions (without corrosion and with corrosion) of steel 12X18H10T (heat exchange tube) was performed. According to the results of microanalysis in the vicinity of corrosive ulcers, the chemical composition of the main elements was as follows (% wt.): 72.4 Fe; 15.2 Cr; 10.8 Ni; 1.1 Ti, which practically corresponds to the composition of steel 12X18H10T, except for the tendency to decrease the concentration of chromium. However, as the authors [4] showed, the chemical composition of steel has changed radically in the ulcer itself, where locally, along with the increase in carbon concentration, the concentration of chromium has increased significantly. Excess carbon content causes the formation of chromium carbides and depletion of the surrounding austenite with chromium with a concentration below 12%, which no longer ensures the corrosion resistance of steel.

Other authors [10] showed that the level of concentration fluctuations of chemical elements from the boundary into the grain grain of austenitized steel

12X18H10T decreases: for nickel—~1.5 times its average content in steel, chromium—20% of the average content, for sulfur and phosphorus—dozens of times. Therefore, it should be expected that in industrial austenitic chromium-nickel steels the corrosion is local and depends on the atomic-magnetic structure of austenite, the content of which in the studied steels ~99.5% (δ-ferrite not more than 0.5%). Many works are devoted to the ambiguous (sometimes opposite) influence of the α-phase (δ-ferrite, α′-martensite) on the corrosion resistance of austenitic chromium-nickel steels [11]. At the content of 0.06...0.08% of δ-ferrite in the austenitic paramagnetic matrix, the magnetic moments of δ-ferrite and austenite are equal. Therefore, to more accurately determine the amount of δ-ferrite in the sample by magnetometric method, the magnetization of the austenitic matrix was taken into account.

In **Figure 2** shows the relationship between the corrosion rate K and the content of P_δ δ-ferrite in steels AISI 304 and AISI 321. As you can see, the amount of δ-ferrite can determine the corrosion rate. However, a significant effect of this amount of P_δ δ-ferrite (0.005...0.12% in steels AISI 304, 08X18H10 and 0.01...0.30% in steels AISI 321, 08X18H10T) is unlikely due to insignificant (compared with austenite) the contact surface of δ-ferrite with an aggressive environment.

To further elucidate the role of low δ-ferrite content on the corrosion process, an experimental dependence of P_δ on χ_0 was constructed (**Figures 3** and **4**).

The low amount of P_δ δ-ferrite correlates with the specific paramagnetic susceptibility χ_0 of austenite (the amount of which is ~99.7%), and hence with the atomic-magnetic (paramagnetic) state of austenite (χ_0). In the studied steels, the amount of δ-ferrite is in thermodynamic equilibrium with the paramagnetic state of austenite, the smallest violation of which causes a change in the amount of δ-ferrite in the austenitic matrix. Hence there is an indirect dependence of the corrosion rate K on the content of P_δ δ-ferrite, which, in turn, corresponds to the atomic-magnetic state of austenite (parameter χ_0). Therefore, it is assumed that the low content of δ-ferrite indirectly (not directly) affects corrosion, i.e. is a measure (indicator) of the corrosion rate K.

It is experimentally established that the corrosion resistance of austenitic chromium-nickel steels depends on the atomic-magnetic state of almost 100% austenite—paramagnet, which is characterized by a specific magnetic susceptibility

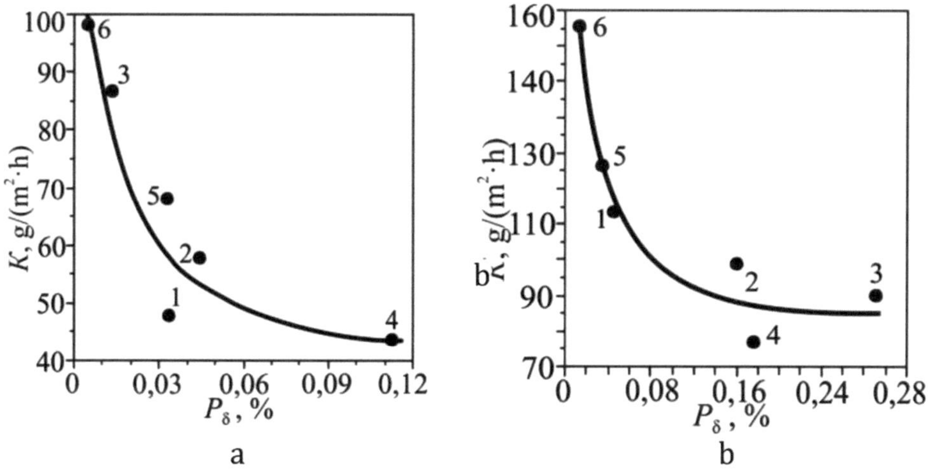

Figure 2.
The relationship between the corrosion rate K *(involved with [9]) and the content of* P_δ *δ-ferrite in the studied steels. a—AISI 304 (melt 1–5) and 08Cr18Ni10 (melt 6); b—AISI 321 (melt 1–5) and 08Cr18Ni10Ti (melt 6).*

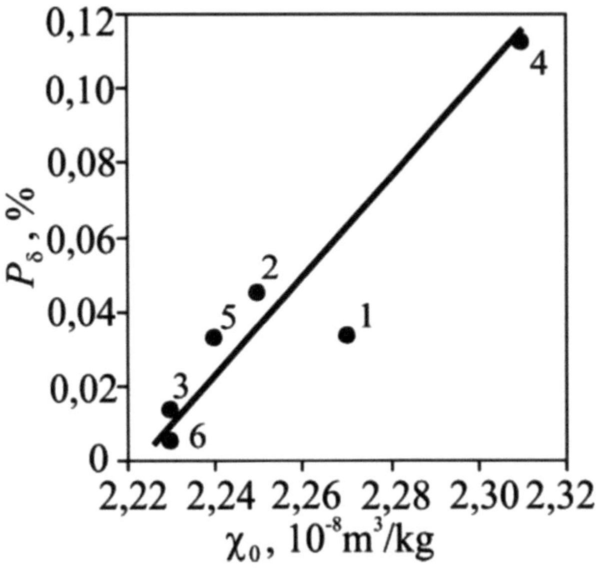

Figure 3.
Tendency of change of low amount of P_δ δ-ferrite from atomic-magnetic state of austenitic matrix (parameter χ_o) in steels AISI 304 (melt 1–5) and 08Cr18Ni10 (melt 6).

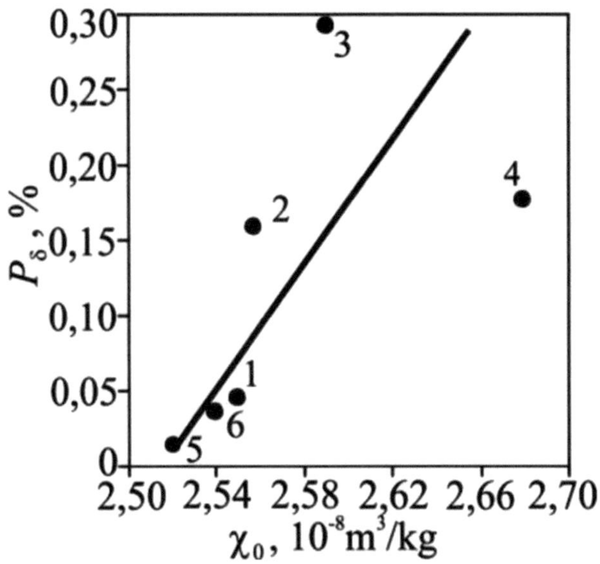

Figure 4.
Tendency of change of low amount of P_δ δ-ferrite from atomic-magnetic state of austenitic matrix (parameter χ_o) in steels AISI 321 (melt 1–5) and 08Cr18Ni10Ti (melt 6).

χ_0: the greater χ_0, the higher the corrosion resistance (lower corrosion rate K), and vice versa. It is assumed that the low content of P_δ δ-ferrite (0.005...0.5%) is an indicator of corrosion intensity, because P_δ depends on χ_0 (with increasing χ_0 austenite increases P_δ), and K depends on χ_0 (with increasing χ_0 decreases K). It should be noted that as the amount of P_δ δ-ferrite increases, the magnetic susceptibility χ of steel (rather than χ_0 austenite), which contains both austenite and δ-ferrite, increases. It is shown that with increasing P_δ (at low δ-ferrite content) the corrosion rate K decreases (and vice versa), i.e. low δ-ferrite content can also be a measure (indicator) of the corrosion process in austenitic chromium-nickel steels.

3. Investigation of corrosion resistance of different melts of one brand of austenitic Fe-Cr-Ni alloys that do not contain δ-ferrite

Considerable attention has been paid to the study of the influence of various factors on the corrosion process of austenitic chromium-nickel steels. In the previous section, the dependence of the corrosion rate K of AISI 304, 08Cr18Ni10, AISI 321, 08Cr18Ni10Ti steels containing δ-ferrite on the atomic-magnetic state of austenite, i.e. on the specific paramagnetic susceptibility χ_0 of austenite, which is an integral value and many factors on corrosion behavior. In this regard, it is necessary to investigate the corrosion of austenitic Fe-Cr-Ni alloys that do not contain α-phase. Five melts of 08Cr28N27 alloy, which does not contain δ-ferrite, were selected. The chemical composition of the melts of this alloy are given in **Table 3**.

Samples were selected from sheet (thickness 1 mm) industrial supplies of alloy 08Cr28N27. Since the numerical values of the specific magnetic susceptibility across the width of the cold-rolled sheet were locally unevenly distributed, in order to average from different places in a checkerboard pattern from each melting cut 10 samples in the form of parallelepipeds ($\sim 6 \times 4 \times 1$ mm^3). The obtained values of the specific paramagnetic susceptibility to corrosion tests are given in **Table 4**. For melts 1...5 of alloy 08Cr28Ni27 the average values of specific paramagnetic susceptibility χ_0 of austenite are received: $2.95 \cdot 10^{-8}$; $2.86 \cdot 10^{-8}$; $3.58 \cdot 10^{-8}$; $3.09 \cdot 10^{-8}$; $2.96 \cdot 10^{-8}$ m^3/kg, respectively [12].

In order to accelerate chemical corrosion, a model aggressive medium was used: a mixture of concentrated acids—hydrochloric and nitric ($HCl:HNO_3$ 3:1) and the samples were kept continuously for 0.5 h at t = 30°C. The corrosion rate K was determined by the formula $K = \Delta m/(S \cdot \tau)$, where Δm is the loss of mass before and after corrosion, S is the surface area of the sample, τ is the exposure time in an aggressive environment. **Table 5** shows the obtained average values of corrosion rate: 1381, 1397, 1519, 1540, 1470 g/(m^2 h), respectively, for melts 1...5.

Analysis of the experimental dependences of the corrosion rate K on the paramagnetic susceptibility χ_0 of austenite (to corrosion tests) of different melts, but one grade of alloy 08Cr28N27, which does not contain δ-ferrite (**Figure 5**) shows: the greater χ_0, the greater the corrosion rate K [12], i.e. there is an opposite dependence compared to austenitic chromium-nickel steels that contain δ-ferrite (see **Figure 1**).

Therefore, the selected sensitive parameter χ_0, able to distinguish the corrosion rate of similar chemical composition of different melts of the same brand of alloy 08Cr28Ni27. For the existence of a paramagnetic effect, it is necessary that the electronic shells of matter have uncompensated orbital and spin magnetic moments, which are oriented in the magnetic field H. Hence, by studying the magnetic

Melt №	Element, % wt.									
	C	Mn	Si	Cr	Ni	Ti	S	P	Mo	Cu
1	0.050	0.34	0.49	24.29	27.40	0.80	0.006	0.028	2.88	2.73
2	0.067	0.43	0.56	22.69	27.62	0.60	0.005	0.029	2.77	2.68
3	0.068	0.50	0.55	21.82	27.47	0.57	0.004	0.032	2.56	2.62
4	0.048	0.55	0.71	22.69	27.70	0.67	0.006	0.028	2.56	2.54
5	0.050	0.29	0.57	23.44	27.48	0.90	0.004	0.032	2.53	2.76

Table 3.
Chemical composition of melts of 08Cr28Ni27 alloy.

№ sample	χ_o, 10^{-8} m^3/kg				
	Melting number				
	№1	№2	№3	№4	№5
1	2.99	2.85	3.37	3.05	2.94
2	2.94	2.83	3.42	3.05	2.92
3	3.06	2.86	3.32	2.92	3.03
4	3.01	2.88	3.33	3.11	2.96
5	2.96	2.85	3.44	3.12	2.93
6	2.95	2.89	3.35	3.09	2.95
7	2.91	2.88	3.40	3.10	3.00
8	2.89	2.81	3.38	3.13	3.04
9	2.92	2.90	3.41	3.16	2.91
10	2.90	2.87	3.43	3.12	2.94
Average	2.95	2.86	3.38	3.09	2.96
$\pm\Delta_{abs.}$	0.04	0.02	0.04	0.05	0.04
$\pm\Delta_{rel.}$,%	1.41	0.77	1.03	1.52	1.24

Table 4.
The value of the specific magnetic susceptibility χ_o of austenite cut samples from the melts № 1…5 alloy 08Cr28Ni27.

№ sample	*K*, g/(m^2•h)				
	Melting number				
	№1	№2	№3	№4	№5
1	1385	1385	1541	1545	1486
2	1420	1439	1508	1510	1467
3	1355	1366	1510	1555	1451
4	1388	1365	1498	1560	1485
5	1313	1359	1494	1570	1464
6	1398	1466	1528	1569	1447
7	1385	1432	1507	1525	1485
8	1378	1388	1527	1505	1453
9	1402	1369	1529	1549	1485
10	1388	1399	1546	1510	1479
Average	1381	1397	1519	1540	1470
$\pm\Delta_{abs.}$	19.5	29.7	15.4	21.8	13.8
$\pm\Delta_{rel.}$,%	1.41	2.13	1.01	1.41	0.93

Table 5.
The value of corrosion rate K *cut samples of the alloy 08Cr28Ni27.*

properties of austenite, it is possible to obtain information about the behavior of austenite under the influence of external factors, such as its corrosion resistance.

The obtained results do not contradict the value of the corrosion rate K (attracted from [9]) for these 1…5 melts of alloy 08Cr28Ni27 (respectively: 0.095;

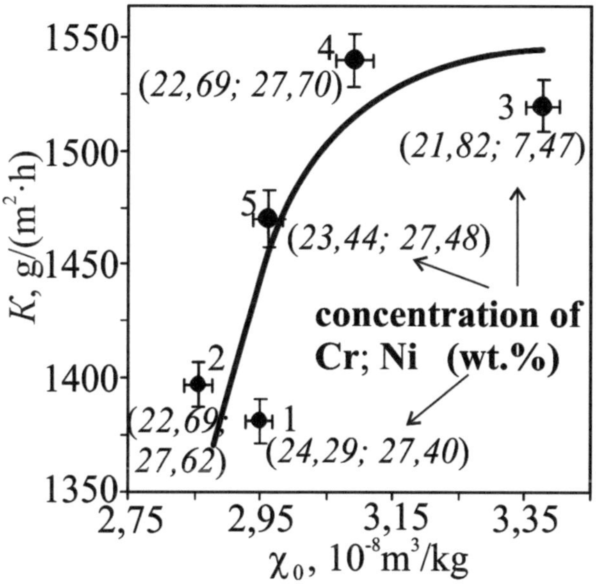

Figure 5.
Tendencies of change of corrosion rate K *in a mixture of concentrated acids-hydrochloric and nitric ($HCl:HNO_3$ 3:1) and specific magnetic susceptibility χ_0 austenite of different melts of alloy 08Cr28Ni27, which do not contain δ-ferrite.*

0.095; 0.143; 0.190 and 0.143 g/(m^2 h)). In the environment that contains chlorine (**Figure 6**).

Compare the graphs of **Figures 5** and **6**. From **Figure 6** it follows that the corrosion rates for two pairs of swimming trunks 1, 2 and 3, 5 are the same (0.095 and 0.143 g/(m^2 h)), and according to our studies, the corrosion rates of all these

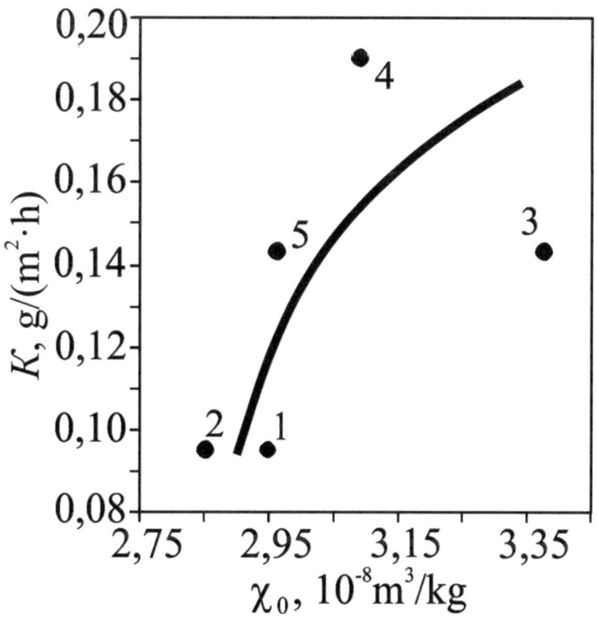

Figure 6.
The relationship of borrowed values of the corrosion rate K *[9] in a chloride-containing medium and found the specific magnetic susceptibility χ_0 [12] of austenite of the same melts of the alloy 08Cr28Ni27, which do not contain δ-ferrite.*

five swimming trunks are different (which corresponds to different χ_0), i.e. the resolution is slightly better, but coincides qualitatively with the general nature of the curves (**Figures 5** and **6**).

It is experimentally established that the corrosion resistance of the austenitic alloy 08Cr28Ni27 (analog AISI 904L) correlates with the atomic-magnetic state of austenite, which is determined by the susceptibility χ_0: the less χ_0, the higher the corrosion resistance (lower corrosion rate), in contrast to austenitic chromium-nickel steels, which contain δ-ferrite. The obtained results can be used to predict the local corrosion resistance of austenitic alloys that do not contain δ-ferrite.

4. Magnetometric evaluation of corrosion resistance of austenitic chromium-nickel steels depending on the phase composition (*A, F, M*) after deformation by compression and bending of specimens

Despite numerous studies of corrosion processes in stainless austenitic steels, which are widely used in aviation and nuclear energy, questions concerning the nature of corrosion still remain unclear. Many works are devoted to the influence of α-phase (δ-ferrite, α′-martensite deformation) on the corrosion resistance of steels [4, 5, 13, 14]. However, the corrosion resistance of austenite (*A*), δ-ferrite (*F*) and α′-martensite deformation (*M*) and the phases $A + F$, $A + M$, $A + F + M$ present together are insufficiently studied. In [13] it was found that the significant cause of corrosion resistance of austenitic steels is not the low content of the α-phase, but the atomic-magnetic state of the austenitic matrix, which is determined by the specific magnetic susceptibility.

To clarify the role of paramagnetic austenite (*A*), δ-ferrite (*F*), α′-martensite deformation (*M*) in the individual and total influence of the phases $A + M$, $A + F$, $A + F + M$ on the corrosion resistance of chromium-nickel steels austenitic class were selected two industrial stainless steels (№1 and №2) of one grade 08Cr18Ni10Ti (sheets with a thickness of 1 mm) with a slight difference in the content of Ni and other elements.

Chemical composition of steels (% wt.): №1—0.08 C; 17.74 Cr; 10.56 Ni; 0.259 Ti; 0.982 Mn; 0.23 Si; 0.04 S and №2—0.09 C; 18.2 Cr; 10.1 Ni; 0.56 Ti; 0.75 Mn; 0.7 Si; 0.01 S; 0.026 P; 0.14 Cu; 0.05 Co; 0.04 V; 0.04 W; 0.06 Mo.

The samples were cut by cold mechanical method in the form of rectangular parallelepipeds of approximately the same size $7 \times 3 \times 1$ mm^3. The degree of residual deformation *D* was calculated by the ratio of thickness before (d_0) and after (d) deformation ($D = (d_0 – d)/d_0$). After compression deformation, the samples were tested for corrosion. In order to accelerate chemical corrosion, the test samples were placed in a mixture of concentrated acids-hydrochloric and nitric ($HCl{:}HNO_3$ 3:1) and kept for 1 h. To visually detect changes in the corrosion rate, we used the coefficient of intense corrosion *K*, which was defined as the relative mass loss $K = (\Delta m/(m \cdot \tau)) \cdot 100\%$.

Austenitization of steel №1 was performed in the standard way (annealing at 1050°C, holding for 30 min, followed by quenching in water). After such heat treatment, the steel №1 in the initial state was paramagnetic, ie single-phase (*A*) with a specific magnetic susceptibility of austenite $\chi_0 = 2.81 \cdot 10^{-8}$ m^3/kg.

Samples of №2 steel containing δ-ferrite in the initial state were not subjected to austenitization, i.e. in the initial state it was two-phase ($A + F$), which were tested after bending at an angle of 180°. After testing the samples with a length of 7 mm per bend, the amount of deformation martensite was distributed unevenly from the ends to the ribs. Therefore, the average value of the amount of α′-martensite for the whole sample was approximated. Due to the fact that the low content of δ-ferrite is unevenly distributed over the width of the cold-rolled sheet in steel №2 [5] and in order to further average the results from different places were cut three, adjacent samples, which after averaging were assigned numbers 1...6.

Steel №1 was tested under uniaxial compression, and steel №2—under deformation by bending.

4.1 Corrosion resistance of austenite of single-phase (A) steel 08Cr18Ni10Ti (№1) at uniaxial compression of specimens to the formation of deformation martensite

The effect of "growth" of the paramagnetic state, i.e. increase of the specific magnetic susceptibility of austenite χ_0 (A) from $2.81 \cdot 10^{-8}$ m^3/kg to the maximum value of $3.2 \cdot 10^{-8}$ m^3/kg of steel №1 at uniaxial compression D from 0 to 4.85%. With a further increase in deformation (accumulation of α′-martensite) from D_S = 4.85% (true martensite point [7]), the specific magnetic susceptibility of austenite did not change and remained equal to $\chi_0^{max} = 3.2 \cdot 10^{-8}$ m^3/kg = const. **Table 6** shows the results of experimental studies of steel №1 (at 20°C) [15].

In **Figure 7** (figures indicate the sample numbers from **Table 6**) shows the change in the corrosion rate $K(A)$ of austenite samples of single-phase (A) steel №1, deformed before the formation of α′-martensite, from changes in the atomic-magnetic state, which is characterized by specific magnetic susceptibility χ_0 (A) austenite.

With uniaxial compression of austenitic specimens, the corrosion rate $K(A)$ increases with the "increase" of the atomic-magnetic state ($\chi_0(A)$), i.e., with an

№ sample	D, %	$\chi_0(A)$, 10^{-8} m^3/kg	χ_0, 10^{-8} m^3/kg	$P_{\alpha'}$, %	$K(A)$, %/h	$K(A+M)$, %/h	$K(M)$, %/h
1	0	2.81	2.81	0	16.80	16.80	0
2	0.90	2.82	2.82	0	17.30	17.30	0
3	1.30	2.83	2.83	0	18.10	18.10	0
4	1.41	2.84	2.84	0	18.20	18.20	0
5	1.52	2.91	2.91	0	20.42	20.42	0
6	2.70	3.01	3.01	0	22.14	22.14	0
7	4.73	3.09	3.09	0	24.03	24.03	0
8	6.51	3.20	3.15	0.02	24.1	24.45	0.35
9	7.35	3.20	3.81	0.05	24.1	24.95	0.85
10	8.15	3.20	3.99	0.14	24.1	25.02	0.92
11	9.95	3.20	4.27	0.19	24.1	25.71	1.61
12	12.43	3.20	6.72	0.45	24.1	27.40	3.30
13	19.13	3.20	10.52	0.79	24.1	29.91	5.81
14	26.17	3.20	18.11	2.18	24.1	33.35	9.25
15	36.0	3.20	27.9	3.14	24.1	36.50	12.40

Table 6.
The value of the amount of ferrophase $P_{\alpha'}$*, the specific magnetic susceptibility* χ_0(A)*, the resulting magnetic susceptibility* χ_∞ *of the sample and the corrosion rate* K *after deformation by compression of steel samples №1, 08Cr18Ni10Ti.*

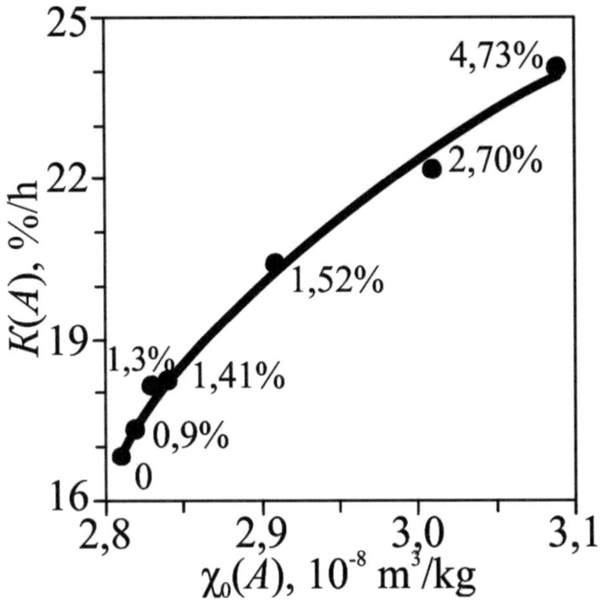

Figure 7.
Change in the corrosion rate K(A) of austenite steel samples 08Cr18Ni10Ti (№1), deformed by compression (degree of deformation—number near points) to the emergence of α′-martensite deformation ($\gamma \to \gamma'$), from changes in the magnetic state of austenite (χ_o).

increase in the specific magnetic susceptibility $\chi_0(A)$ of austenite, the corrosion rate $K(A)$ increases (corrosion resistance decreases).

4.2 Corrosion resistance of two-phase ($A + M$) steel 08Cr18Ni10Ti (№1) at uniaxial compression (occurrence of α′-martensite)

With increasing degree of plastic deformation compression after the actual deformation point $D_S = 4.85\%$ occurs and accumulates α′-martensite deformation, i.e. the continuation of the curve (see **Figure** 7) is the curve after D_S, which is presented in **Figure** 8. Hence we have that the corrosion rate of two-phase ($A + M$) steel №1 increases with increasing plastic deformation. As indicated, after deformation $D_S = 4.85\%$ and above, the value of $\chi_0{}^{max}(A) = 3.2 \cdot 10^{-8}$ m^3/kg remains constant, i.e. $\chi_0{}^{max}(A) = const$. Therefore, the corrosion of austenite after this point (D_S) is constant. Therefore, the corrosion rate of steel 08Cr18Ni10Ti (№ 1) in the area $D = 0...4.85\%$ increases due to the deformation of austenite, and in the area $D \geq 4.85\%$ increases due to corrosion of α′-martensite deformation that accumulates when a constant value of the corrosion rate of austenite [15].

The dependence of the corrosion rate $K(A + M)$ on the amount of α′-martensite deformation $P_{\alpha'}$ occurring during steel compression №1 is shown in **Figure 9**. As we can see, the corrosion rate increases with increasing amount of α′-martensite $P_{\alpha'}$.

4.3 Corrosion resistance of deformation α′-martensite (M), which occurs during uniaxial compression of steel 08Cr18Ni10Ti (№1)

The maximum value of the specific paramagnetic susceptibility of austenite $\chi_0{}^{max}(A) = 3.2 \cdot 10^{-8}$ m^3/kg $= const$ corresponds to $K(A) \approx 24\%$/h (see **Figure 8**). If we subtract this value of the corrosion rate of austenite $K(A)$ from the final

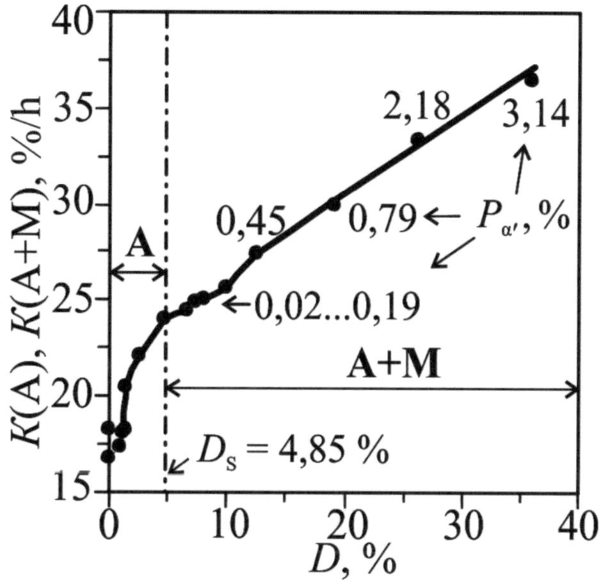

Figure 8.
Change in the corrosion rate of steel samples 08Cr18Ni10Ti (№1) during the transition from single-phase state K(A) *to two-phase state* K(A + M) *(amount of α′-martensite deformation—number near points).* D_S*—real deformation martensitic point (beginning* $\gamma \rightarrow \alpha'$ *transformation).*

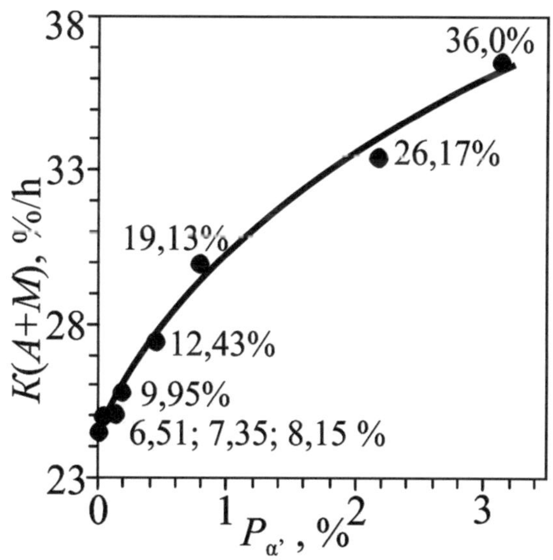

Figure 9.
Change in the corrosion rate K(A + M) *from the content of α′-martensite deformation under uniaxial compression (degree of deformation—number near the points) of steel samples 08Cr18Ni10Ti (№1).*

corrosion rate $K(A + M)$ (see **Figure 9**), we obtain the value of the corrosion rate $K(M)$ from the amount of α′-martensite (**Figure 10**). Hence, we have that the corrosion rate $K(\mathrm{M})$ increases with increasing amount of $P_{\alpha'}$ α′-martensite deformation.

In **Figure 11** presents a model of corrosion processes for single-phase (separately austenite (A) and α′-martensite deformation (M)) and two-phase ($A + M$) states of steel 08Cr18Ni10Ti (№1) [15].

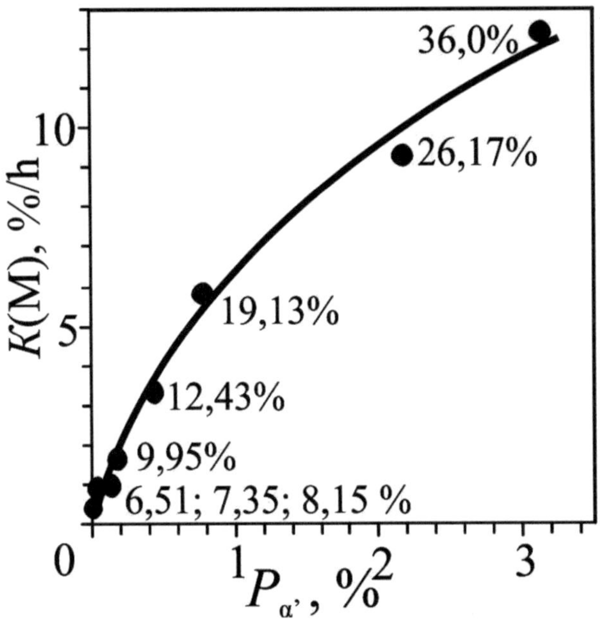

Figure 10.
Change of corrosion rate K(M) *from the content of α′-martensite deformation* $P_{\alpha'}$ *under uniaxial compression (degree of deformation—number near points) of steel samples 08Cr18Ni10Ti (№1).*

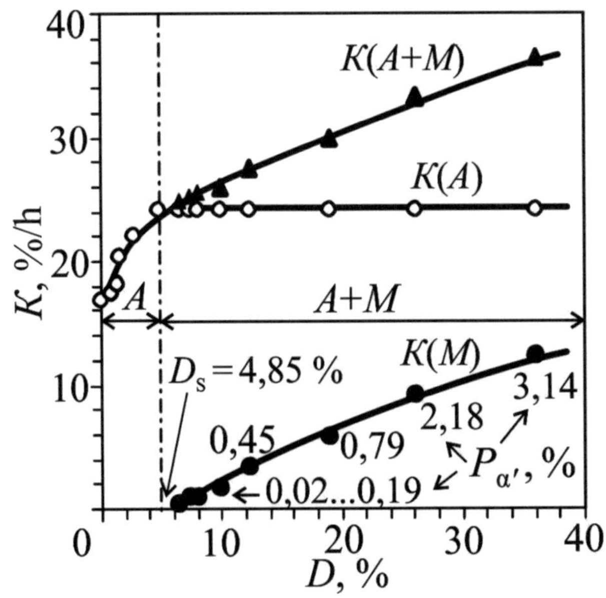

Figure 11.
Change of corrosion rate K(A + M), K(M), K(A) *from the degree of deformation* D *by compression of steel samples 08Cr18Ni10Ti (№1).* D_S*—real deformation martensitic point (beginning* $\gamma \to \alpha'$ *transformation);* $P_{\alpha'}$*—amount of α′-martensite deformation (numbers near the points of the curve* K(M)*).*

4.4 Corrosion resistance of three-phase ($A + F + M$) steel 08Cr18Ni10Ti (№ 2) after bending of samples at an angle of 180° from the initial atomic-magnetic state ($\chi_0(F + M)$) of austenitic matrix containing δ-ferrite and α′-martensite deformation

From the industrial sheet steel №2 were cut from different cities on 3 adjacent samples with the subsequent averaging of the measured values (amount of δ-ferrite, α′-martensite, corrosion rate K, specific magnetic susceptibility χ_0 austenite). A total of 18 samples were cut, from which 6 averaged samples were grouped, i.e. 6 "points". The average values of the amount of δ-ferrite and the specific magnetic susceptibility of $\chi_0(F)$ austenite in the presence of δ-ferrite were measured before bending deformation, and the amount of total δ-ferrite and α′-martensite after bending deformation (180°).

Six samples were obtained with averaged values of P_δ before deformation and $P_{\delta+\alpha'}(180)$ after bending at an angle of 180°. For each of the three neighboring samples according to the method determined $\chi_\infty(F)$ and from the graphical dependence $\chi_\infty(F) = f(P_\delta)$ by extrapolation ($P_\delta \to 0$) found the specific paramagnetic susceptibility of the austenitic matrix $\chi_0(F)$, which contains δ-ferrite. The results obtained are given in **Table 7**. According to the literature, the amount of δ-ferrite P_δ during deformation does not change and subtracting it from the experimentally found values of $P_{\delta + \alpha'}(180)$, found the amount of formed α′-martensite deformation $P_{\alpha'}(180)$ (**Table 7**).

From the graphical dependence of $\chi_\infty(F + M)$ on $P_{\delta + \alpha'}$ (by analogy with steel №1) by the method of extrapolation, when $P_{\delta + \alpha'} \to 0$ (both ferrophases are absent) found the value of the specific magnetic susceptibility of austenite $\chi_0^{max}(A) = 2.2 \cdot 10^{-8}$ m³/kg = *const* (for austenitic matrix of steel №2, which contained δ-ferrite (F) and α′-martensite (M)).

In **Figure 12** [15] presents the change in the corrosion rate $K(A + F + M)$ of three-phase ($A + F + M$) steel №2 after bending the samples at an angle of 180° from the magnetic state ($\chi_0(F + M)$) of the austenitic matrix, which contains δ-ferrite and α′-martensite.

From **Figure 12** we have that the corrosion rate of three-phase ($A + F + M$) steel №2 after bending of two-phase ($A + F$) samples by an angle of 180° increases with increasing specific magnetic susceptibility $\chi_0(F + M)$ of the original samples with δ-ferrite. Thus, the content of phases ($A + F + M$) increases the corrosion rate. A similar dependence of $K(A + F + M)$ on $P_{\delta + \alpha'}$ is shown in **Figure 13**.

№ sample	P_δ, %	$P_{\delta + \alpha'}(180)$, %	$P_{\alpha'}(180)$, %	$\chi_0(F)$, 10^{-8} m³/kg	$K(A + F + M)$, %/h
1	0.04	0.87	0.83	1.9	12.2
2	0.27	1.47	1.20	3.6	18.0
3	0.12	1.07	0.95	2.9	19.5
4	0.09	1.35	1.26	2.5	17.0
5	0.07	0.90	0.83	2.35	15.5
6	0.26	1.29	1.03	3.5	18.3

Table 7.
The value of the amount of ferrophase, the specific magnetic susceptibility χ_0(F) of the original samples and the corrosion rate K(A + F + M) after deformation by bending at an angle of 180° steel samples №2.

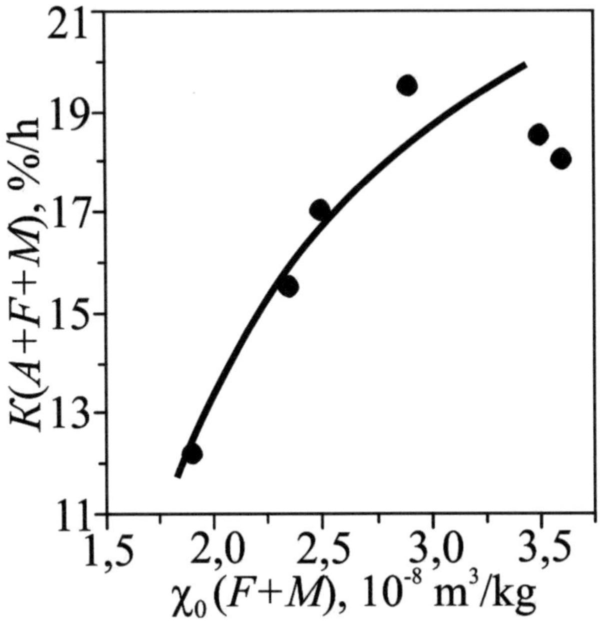

Figure 12.
Relationship between the corrosion rate K(A + F + M) *of three-phase* (A + F + M) *steel 08Cr18Ni10Ti (№2) after bending two-phase* (A + F) *samples at an angle of 180° and the atomic-magnetic state* (χ_o(F + M)) *austenitic matrix of samples containing δ-ferrite* (F) *and α′-martensite deformation* (M).

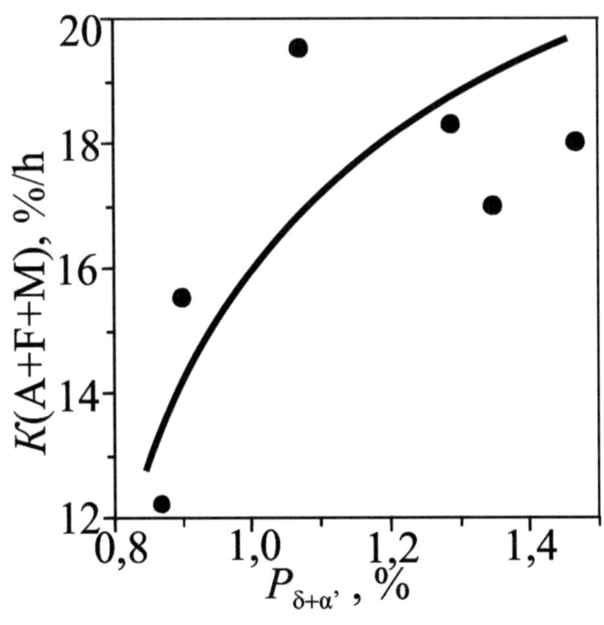

Figure 13.
Relationship between the corrosion rate K(A + F + M) *of three-phase* (A + F + M) *steel 08Cr18Ni10Ti (№2) after bending the two-phase* (A + F) *samples at an angle of 180° and the total number of* $P_{\delta + \alpha'}$ *samples containing δ-ferrite* (F) *and α′-martensite deformation* (M).

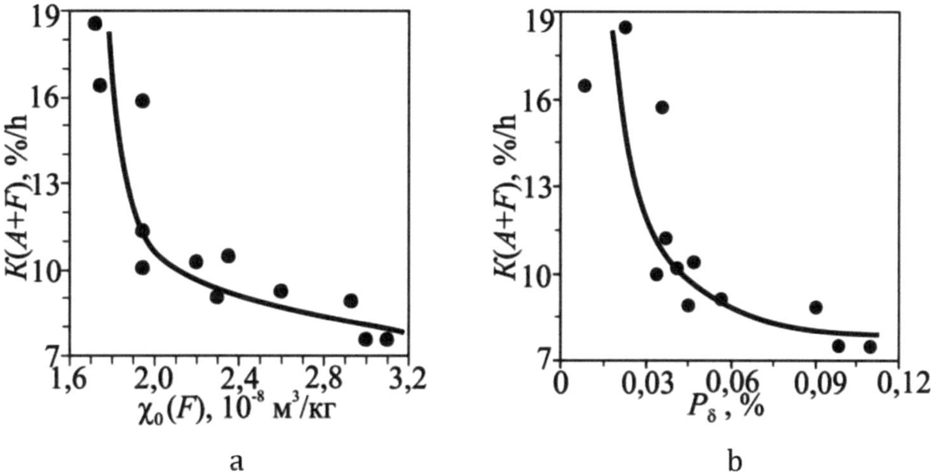

Figure 14.
Change in the corrosion rate K(A + F) *from the specific magnetic susceptibility* χ_0(F) *(a) of the austenitic matrix of samples (containing δ-ferrite) and the content of* P_δ *δ-ferrite (b) steel 08Cr18Ni10Ti (№2).*

4.5 Corrosion resistance of two-phase (*A* + *F*) steel 08Cr18Ni10Ti (№2) to bending deformation depending on the initial atomic-magnetic state ($\chi_0(F)$) of austenitic matrix containing δ-ferrite

To systematize the magnetometric assessment of the corrosion resistance of all types of phases, we use two graphical dependences $K(A + F)$ on $\chi_0(F)$ and $K(A + F)$ on P_δ shown in **Figure 14** [13].

From **Figure 14a** it follows that this dependence has the following character: with the "increase" of the atomic-magnetic state of the austenitic matrix ($\chi_0(F)$) the corrosion rate decreases. Similarly, for the dependence of $K(A + F)$ on P_δ (**Figure 14b**): with increasing amount of δ-ferrite, the corrosion rate decreases (corrosion resistance increases) (**Figure 15**).

4.6 Corrosion resistance of deformation martensite after bending of steel samples 08Cr18Ni10Ti (№2) at an angle of 180° from the initial atomic-magnetic state ($\chi_0(F)$) of the austenitic matrix containing δ-ferrite

If we subtract the value of the corrosion rate $K(A + F)$ (see **Figure 14a**) from the final corrosion rate $K(A + F + M)$ (see **Figure 15**), we obtain a change in the corrosion rate $K(M)$ of α′-martensite from atomic-magnetic state ($\chi_0(F)$) of the austenitic matrix containing δ-ferrite [15]. In **Figure 16** presents a model of corrosion processes for three-phase ($A + F + M$), two-phase ($A + F$) and single-phase (M) states of steel 08Cr18Ni10Ti (№2).

4.7 Dependence of corrosion resistance of δ-ferrite of steel 08Cr18Ni10Ti (№2) on the initial atomic-magnetic state ($\chi_0(F)$) of austenitic matrix containing δ-ferrite

Taking into account the tendency of change of corrosion rates $K(A)$ and $K(A + F)$ (see **Figures 7** and **14a**), we can propose the dependence of $K(F)$ on $\chi_0(F)$,

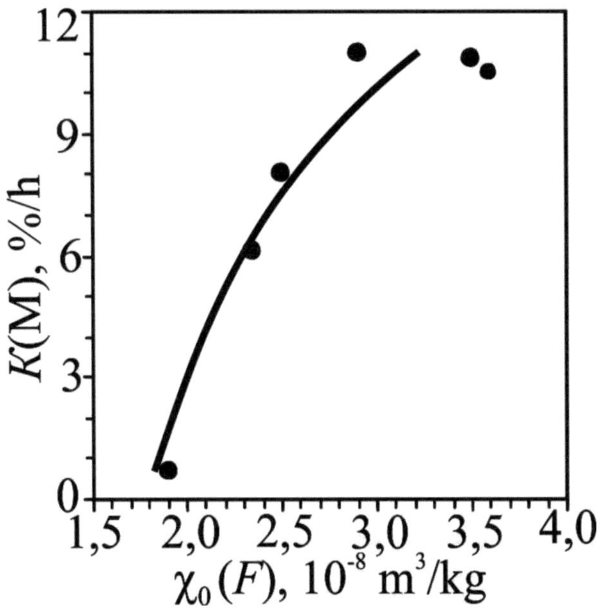

Figure 15.
Change in the corrosion rate K(M) *from the specific magnetic susceptibility* χ_0(F) *of the austenitic matrix containing δ-ferrite, steel 08Cr18Ni10Ti (№2).*

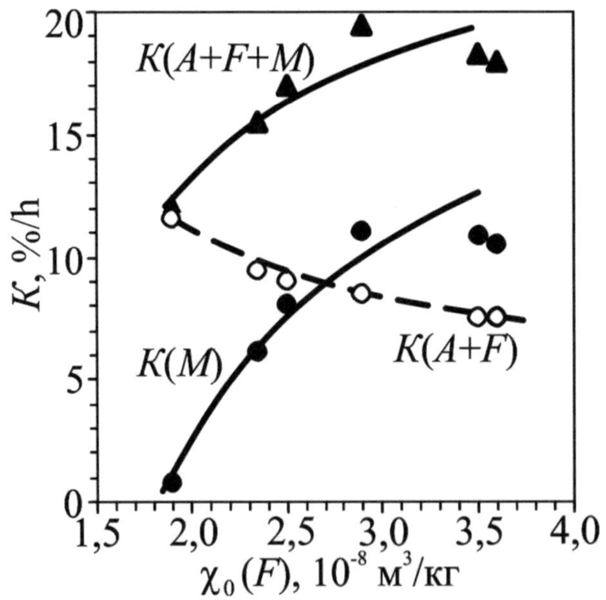

Figure 16.
Change of corrosion rate K(A + F + M), K(A + F) *and* K(M) *from the specific magnetic susceptibility* χ_0(F) *of the austenitic matrix containing δ-ferrite, steel 08Cr18Ni10Ti (№2).*

which is presented in **Figure 17**. As we can see, the corrosion rate of $K(F)$ δ-ferrite decreases (corrosion resistance increases) [15].

Based on experiments, it was found that austenite steels 08Cr18Ni10Ti (single-phase state A after austenitization), deformed by compression to the actual deformation martensitic point D_S, acquires an increased corrosion rate $K(A)$. At

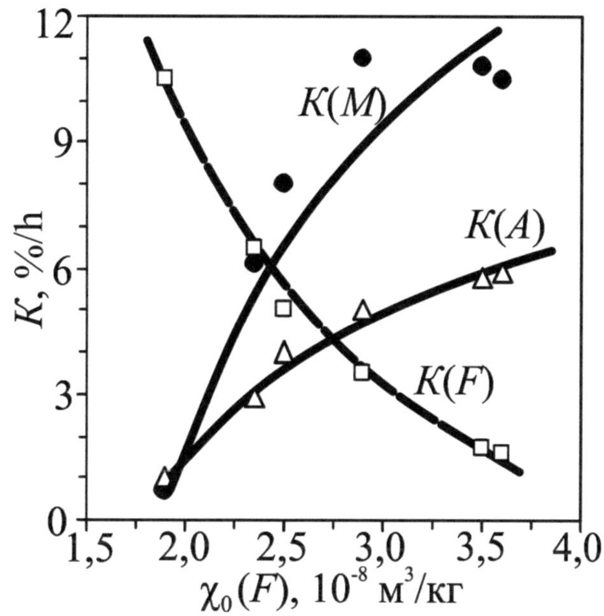

Figure 17.
Change in the corrosion rate of individual phases: K(A) *austenite,* K(F) *δ-ferrite and* K(M) *α′-martensite after deformation by bending from the initial atomic-magnetic state of the austenitic matrix (parameter χ_o(F)), which contains δ-ferrite, steel samples 08Cr18Ni10Ti (№2).*

deformations D_S and above, the corrosion rate $K(A)$ of austenite remains constant, and the corrosion rate $K(M)$ α′-martensite increases.

The corrosion rates of $K(F)$ δ-ferrite and $K(M)$ α′-martensite of the studied steels are opposite: $K(F)$ decreases, and $K(M)$ increases with increasing deformation, i.e. with increasing atomic-magnetic state of austenite ($\chi_0(F)$).

Corrosion rates $K(A + F)$ and $K(A + M)$ of the studied steels in accordance with the two-phase states $A + F$ and $A + M$ have the opposite tendency: $K(A + F)$ with increasing $\chi_0(F)$ decreases (corrosion resistance increases), and $K(A + M)$ on the contrary—increases.

The corrosion rate $K(A + F + M)$ of three-phase steel increases with deformation, ie with increasing atomic-magnetic state ($\chi_0(F + M)$) of the austenitic matrix, which contains δ-ferrite and α′-martensite deformation.

5. Conclusions

1. It was experimentally established that the corrosion rate K correlates with the specific paramagnetic susceptibility χ_0 of austenite steels AISI 304, 08Cr18Ni10, AISI 321, 08Cr18Ni10Ti, containing a low content of δ-ferrite (~0.005 ... 0.5%): the greater χ_0, the higher the corrosion resistance (lower corrosion rate K) of steel.

2. It was found that the corrosion rate K of austenitic alloys 08Cr28Ni27, which do not contain δ-ferrite, in contrast to austenitic chromium-nickel steels with a low content of δ-ferrite, has the opposite character: the greater χ_0, the less corrosion resistance (more corrosion rate K) of the alloy.

3. Since χ_0 determines the atomic-magnetic structure of austenite, it is assumed that the corrosion behavior of austenitic chromium-nickel steels depends on

the atomic-magnetic (paramagnetic) state of austenite, previously formed before the interaction with the aggressive environment.

4. The direct effect of low content of δ-ferrite (~0.005...0.5%) in austenitic chromium-nickel steels is questionable due to the small (compared to austenite) contact surface with the aggressive environment. An indirect dependence of the corrosion rate K on the low content of P_δ δ-ferrite, which, in turn, depends on the atomic-magnetic (paramagnetic) state of austenite, i.e. on the parameter χ_0. Therefore, it is assumed that the low content of δ-ferrite indirectly affects the corrosion, i.e. is a measure (indicator) of the corrosion rate of K. It is proposed to predict the intensity of corrosion by the parameters χ_0 and P_α.

5. For austenitic chromium-nickel steels the corrosion rate of separate phases is established: austenite (A), ferrite (F), α′-martensite deformation (M) and total phases $A + F$, $A + M$ and $A + F + M$. Thus, the corrosion rates $K(A)$ and $K(M)$ increase with increasing $\chi_0(F)$ of the austenitic matrix containing δ-ferrite. For two-phase steels $A + F$ and $A + M$, respectively, the corrosion rates $K(A + F)$ and $K(A + M)$ have the opposite character, namely $K(A + F)$ with increasing $\chi_0(F)$ decreases, and $K(A + M)$ on the contrary increase. For three-phase steel $A + F + M$, the corrosion rate $K(A + F + M)$ increases after plastic deformation with increasing $\chi_0(F + M)$ of the austenitic matrix, which contains δ-ferrite and α′-martensite deformation.

Author details

Gennadii Snizhnoi
Department Micro- and Nanoelectronics, National University "Zaporizhzhya Polytechnic", Zaporizhzhya, Ukraine

*Address all correspondence to: gvsneg@gmail.com

References

[1] Cwalina K, Demarest C, Gerard A, Scully J. Revisiting the effects of molybdenum and tungsten alloying on corrosion behavior of nickel-chromium alloys in aqueous corrosion. Current Opinion in Solid State and Materials Science. 2019;**23**(3):129-141. DOI: 10.1016/j.cossms.2019.03.002

[2] Wang X, Yang Z, Wang X, Shia O, Xua B, Zhoa C, et al. The influence of copper on the stress corrosion cracking of 304 stainless steel. Applied Surface Science. 2019;**478**:492-498. DOI: / 10.1016/j.apsusc.2019.01.291

[3] Sabooni S, Rashtchi H, Eslami A, Karimzadeh F, Enayati M, Raeissi K, et al. Imani dependence of corrosion properties of AISI 304L stainless steel on the austenite grain size. International Journal of Materials Research. 2017; **108**(7):552-559

[4] Azhazha V, Desnenko V, Ozhigov L, Azhazha G, Svechkarev I, Fedorchenko A. Application of magnetic methods to the study of the evolution of structures in austenitic stainless steels after the long-term operation of power blocks at NPP. Voprosy Atomnoj Nauki i Techniki. 2009;**4**(94):241-246

[5] Snizhnoi G, Mishchenko V, Snizhnoi V. Prediction of the local corrosion resistance by content α-phase of austenitic chromium-nickel steels. Construction, Materials Science, Mechanical Engineering. Series: "Starodubovskye Reading 2012": Scientific Works Collection. 2012;**64**: 409-416

[6] Dobromyslov A, Taluts N. An electron-microscopic study of the deformation structure of the 12Kh18N10T steel after explosive loading in spherical systems. Diagnostics, Resource and Mechanics of materials and structures. 2015;**5**: 109-117

[7] Snizhnoi G, Rasshchupkyna M. Magnetic state of the deformed austenite before and after martensite nucleation in austenitic stainless steels. Journal of Iron and Steel Research, International. 2012;**19**(6):42-46

[8] Snizhnoi G. Dependence of the corrosion behavior of austenitic chromium-nickel steels on the paramagnetic state of austenite. Materials Science. 2013;**49**(3):341-346

[9] Narivskyi O. Corrosion-electrochemical behavior of structural materials for plate heat exchangers working in model waters [Ph.D. Lviv]. Karpenko Physico-mechanical Institute. Publ.; 2009. p. 200

[10] Ermakov B, Malikov S, Solntsev Y. The role of residual welding strains in the decrease in the serviceability of pipelines exploited under low-temperature conditions. Izvestiya SPGUNiPT. 2007;**1**:28-33

[11] Yarovchuk A, Maksimkin O, Tsai K, Ruban S. Pitting corrosion in cold-deformed 12Cr18Ni10Ti stainless steel. Research and Technology Review National Nuclear Center of the Republic of Kazakhstan. 2014;**1**:38-44

[12] Snizhnoi G, Snizhnoi V. Identification of corrosion resistance different batches of the same brand austenitic Fe-Cr-Ni alloys. Construction, Materials Science, Mechanical Engineering. Series: "Starodubovskye Reading 2016": Scientific Works Collection. 2016;**89**:167-171

[13] Snizhnoi G. Role of magnetic state austenite in the formation of corrosion resistence of austenitic chromium-nickel steels. Aerospace Technic and Technology. 2012;**8**(95):141-144

[14] Mishchenko V, Snizhnoi G, Narivs'kyy O. Magnetometric

investigations of corrosion behaviour of AISI 304 steel in chloride-containing environment. Metallofizika i Noveishie Tekhnologii. 2011;**33**(6):769-774

[15] Snizhnoi G. Magnetometric evaluation of corrosion resistance austenitic chromium-nickel steels with different phase composition. Aerospace Technic and Technology. 2013;**9**(106): 195-201

Section 3

Studies on Stainless Steels

Chapter 5

Surface Hardening of Stainless Steel

André Paulo Tschiptschin and Carlos Eduardo Pinedo

Abstract

The addition of nitrogen to stainless steel improves mechanical and corrosion properties. Nitrogen-bearing stainless steel (HNSS) is a new corrosion-resistant alloy class exhibiting better tribological properties. High-pressure and powder metallurgy techniques were developed for the fabrication of HNSS. Solid-state routes allow nitrogen introduction through thermochemical, implantation, or plasma surface treatments. High-temperature gas nitriding (HTGN), carried out in an N_2 atmosphere in the 1000°C range, allows N uptake, obtaining thick, ~0.5–1.0 wt.% N austenitic cases. HTGN is different from conventional nitriding, performed in the 500°C range, where intense Cr_xN_y precipitation occurs, impairing the corrosion resistance. Low-temperature plasma nitriding (LTPN) introduces more N in solution, and colossal supersaturated expanded phases (~45 at.%N) are formed. N supersaturation and compressive stresses increase the hardness of the surface layer to 10–14 GPa. Ferritic, martensitic, duplex, and precipitation-hardened stainless steels can be surface-treated by LTPN, obtaining expanded ferrite and martensite. However, single LTPN stainless steel may prematurely fail when submitted to high loading, as the thin and hard expanded layers collapse due to lack of load-bearing capacity. Duplex-nitriding treatment (HTGN + LTPN) results in a thick nitrogen-rich hardened austenite substrate layer, granting mechanical support and adhesion to the expanded austenite layer.

Keywords: surface hardening, gas nitriding, plasma nitriding, duplex nitriding, HTGN, LTPN, wear resistance

1. Introduction

Since the beginning of the twentieth century, stainless steel has been developed to improve the corrosion resistance of parts in contact with corrosive and oxidative media. These corrosion-resistant alloys have been used in the chemical, petrochemical, automotive, aeronautical, food, medical, and construction industries. Chromium, above 11 wt.%, grants corrosion resistance by forming a nanometric thin and adherent Cr_2O_3 passive layer. When exposed to oxygen, whether in the air or water, this layer prevents corrosion by isolating the alloy from contact with the oxidizing media.

However, chlorine and chlorine ions may damage the passive layer favoring stainless steel's crevice, pitting, and stress corrosion cracking. Mo additions are very effective in improving the resistance to damage of the passive layer by chlorine, although it negatively influences the final price of the stainless steel. On the other hand, N has been thoroughly investigated, since the 1980s, as an alloying element with great potential for protecting the passive layer against damage, being abundant in nature (21 wt.% in the atmosphere), and giving a cost-effective solution for surface protection against corrosion.

Controlled addition of nitrogen to stainless steel has been encouraged over the last three decades due to the possibility of improving the surface properties (not only the corrosion but also the tribological and mechanical properties). High-pressure and powder metallurgy techniques were developed for medium and large-scale fabrication of high nitrogen steels (HNS). Still, in general, these procedures are costly and require sophisticated equipment. Nitrogen-bearing stainless steel is a new class of corrosion-resistant alloys, exhibiting much better surface properties, better corrosion, and wear resistance associated with good bulk mechanical properties: very high strength, good ductility, and toughness. Therefore, considerable emphasis has been placed on liquid and solid-state routes to produce high-performance, low-cost nitrogen-alloyed stainless steels. The liquid state processing routes demand high-pressure metallurgy, which is laborious, demands special equipment, and is costly. In the solid-state production routes, the steel surface and near-surface regions are nitrogen alloyed through thermochemical, implantation, plasma, or laser techniques.

Diffusion surface treatments have been extensively studied and have become, for many applications, current industrial practice. The diffusion of nitrogen and carbon toward the core increases the surface hardness and wear resistance. However, nitrogen and carbon must remain in solid solution. Precipitation of chromium-rich carbides or nitrides reduces the chromium content in the metal matrix, preventing the formation of a continuous passive layer and harming the corrosion resistance of the steel.

2. High-temperature diffusion surface treatment

Berns [1] proposed, in the first half of the 1990s, carrying out a high-temperature nitriding process by exposing austenite to an N_2 atmosphere. Nitrogen dissolves in austenite up to the solubility limit during the high-temperature nitrogen treatment. Nitrogen solubility in austenite is much greater than in the BCC phases. Then, by maintaining stainless steel in a furnace containing a pure N_2 gas atmosphere, the nitrogen equilibrium between the furnace atmosphere and the alloy can be attained. According to Sieverts' law [2], nitrogen can reach contents up to 1 wt.% in solution. Thermocalc [3] phase diagrams considering the N_2 gas phase were calculated to predict the N_2 content in equilibrium in austenite as a function of temperature and partial pressure, as shown in **Figure 1**.

The high-temperature nitriding treatment consists of a case hardening that enriches the stainless steel's surface with nitrogen contents up to 1 wt.%, to a depth of 1–2 mm. Berns [5] named this process solution nitriding (SN). After this pioneer proposal, several research works have followed on studying this solid-state route for introducing high N_2 contents in solution in austenite, being called high-temperature-gas-nitriding—HTGN [6] or high-temperature-solution-nitriding HTSN [7].

The amount of nitrogen dissolved in austenite, in equilibrium with pure N_2 gas atmosphere, increases with decreasing temperature and pressure, as shown in **Figure 2** [6].

Berns [5] envisaged different possibilities of obtaining tailored engineered stainless steels depending on the composition and surface treatment. Therefore, austenitic stainless steels can be HTGN, obtaining a fully austenitic case with an

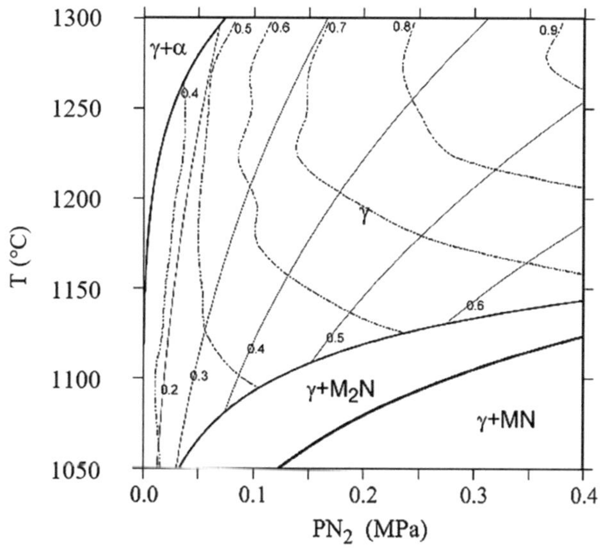

Figure 1.
TPT diagram relating nitriding parameters (nitrogen temperature and partial pressure) with microstructure, nitrogen content, and martensitic layer depth for 3 h gas nitriding treatments at high temperature for an AISI 410S steel [4].

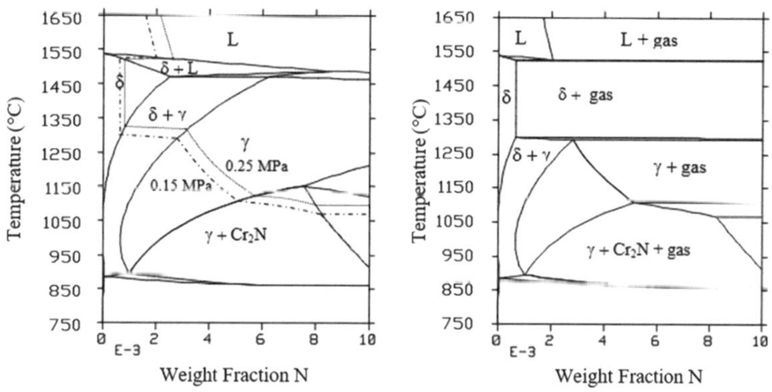

Figure 2.
Fe—13% Cr—N isopleths (a) not considering the gas phase as an equilibrium one, with N_2 isobars overlaid and (b) considering the N_2 gas phase as an equilibrium one [6].

outermost N content of 0.48 wt.%N and case depths of up to 1 mm, with a hardness variation from 1.95 GPa in the low N core to 3.17 GPa in the 0.48 wt.%N case [8]. Martensitic stainless steels can be HTGN, obtaining a much harder 0.4 wt.%N martensitic cases 725 HV hard [9]. Extra-low carbon (0.017 wt.%C) dual-phase stainless steel (α + Martensite) may form a fully martensitic case 550 HV hard, after HTGN [10]. Finally, an UNS 31803 ferritic/austenitic duplex stainless steel can be hardened by HTGN, achieving a fully austenitic layer near the surface due to enrichment in austenite stabilizer element (N), as shown in **Figure 3** [11]. Excess of diffused nitrogen causes a solid solution hardening effect, proportional to its content, reaching a maximum value of 330 HV at maximum concentration, as shown in **Figure 4** [12]. It is worth noting that the N absorption and diffusion on the surface during the HTGN process induce phase transformations, resulting in microstructural gradients from the surface to the core and corresponding microhardness gradients.

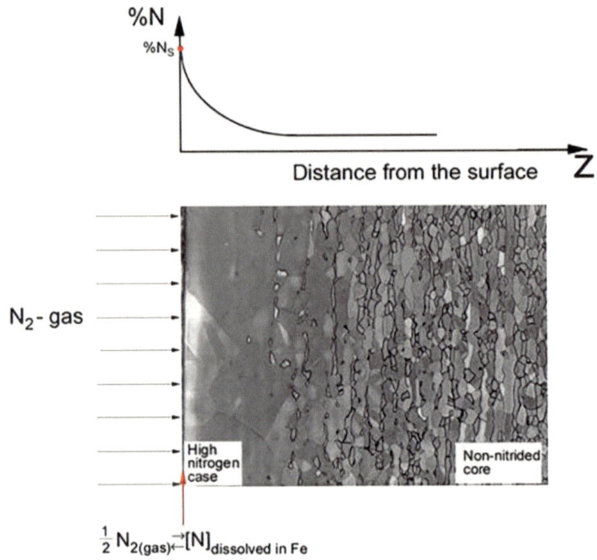

Figure 3.
UNS S31803 duplex stainless steel HTGN at 1200°C [11].

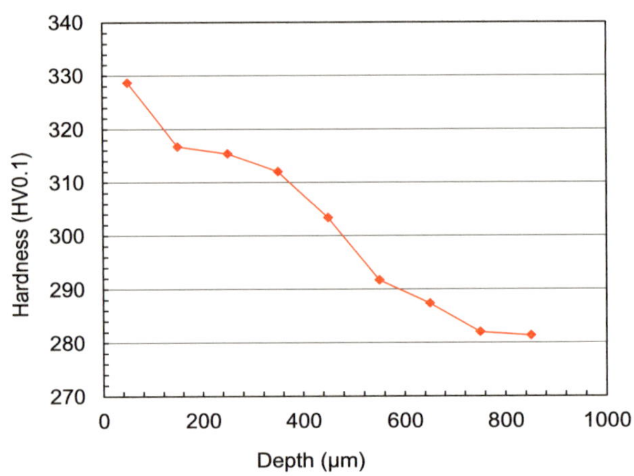

Figure 4.
Microhardness gradient from the low nitrogen duplex ferritic-austenitic core toward the fully austenitic 0.8 wt. %N surface [12].

Tschiptschin [13], using this concept, proposed a Powder Metallurgy route to enrich a ferritic stainless steel powder (0.02 wt.%C, 16.2 wt.%Cr, and 0.81 wt.% Mo), exposing the powder particles at high temperatures (1100°C and 1200°C) to N_2 gas atmosphere. The N enriched austenitic powder transforms during quenching to martensite, becoming very hard. One of the main challenges in this HNS production route is obtaining fully dense components with uniform nitrogen content in volume and excellent surface properties. A uniform nitrogen distribution leads to a more homogeneous microstructure and better mechanical properties. **Figure 5** shows the amount of nitrogen as a function of temperature. According to Sieverts' law [2], increasing temperature decreases the amount of nitrogen content of the obtained alloy. High temperatures are necessary to grant that all the nitrogen is dissolved in austenite, avoiding the precipitation of chromium nitrides.

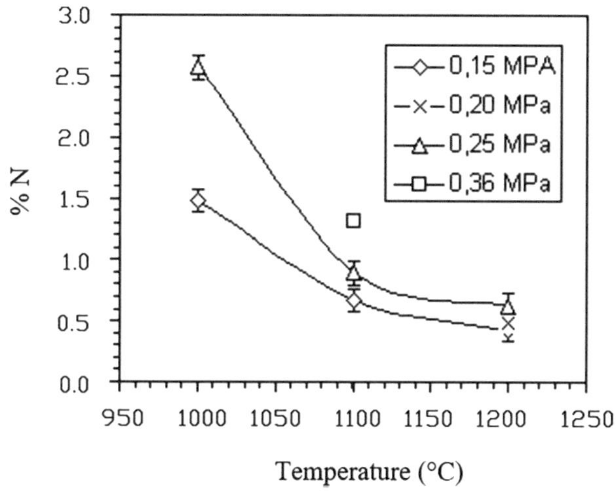

Figure 5.
Nitrogen content as a function of temperature and N_2 pressure for an AISI 434L ferritic stainless steel [13].

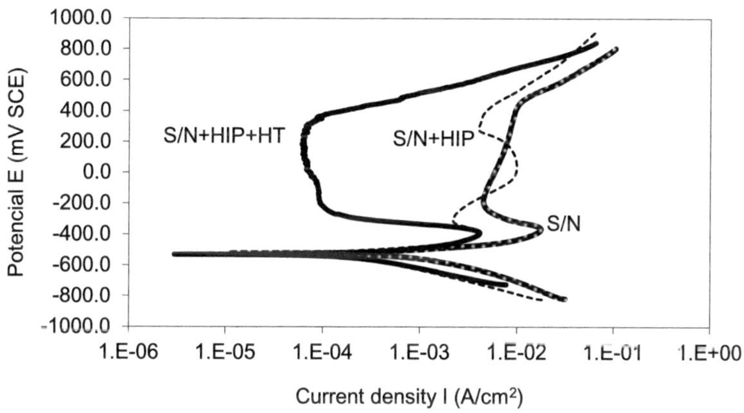

Figure 6.
Cyclic polarization curves for a 0.66 wt.%N martensitic stainless steel in different stages of fabrication. Solution 0.5 M H_2SO_4 + 3.5% NaCl. S/N: Sintered/nitrided, HIP: Hot isostatic pressed, HT: Heat treated (quenched and tempered at 200°C) [13].

In this route, high-nitrogen (0.66 wt.%N) martensitic stainless steel could be obtained by high-temperature gas nitriding an AISI 434L ferritic stainless steel powder, compressing the high nitrogen powder to near net shape parts, followed by hot isostatic pressing and proceeding with a 1200°C quenching and a 200°C tempering treatment. As a result, the obtained hipped material showed high hardness and much better corrosion resistance, measured in potentiodynamic polarization tests carried out in 0.5 M H_2SO_4 + 3.5% NaCl, as shown in **Figure 6**.

3. Low-temperature diffusion surface treatment

Conventional gaseous or liquid nitriding processes are traditionally carried out at temperatures above 520°C. However, this process temperature is a limiting factor, considering that when the nitriding of stainless steels is conducted above 500°C, intense precipitation of chromium nitrides and carbides occurs in the diffusion zone, which, despite substantially increasing the hardness, greatly compromises corrosion resistance [14–18].

The diffusion temperature is the main control parameter to prevent chromium nitrides and chromium carbides precipitation. Precipitation of chromium carbides and nitrides requires substitutional diffusion, which only occurs at temperatures higher than 500°C. Zhang and Bell [14] and Ichii et al. [19] pioneered the study and development of stainless steel's nitrogen and carbon diffusion processes in low temperatures. The process temperature must be selected, not too low, to allow intense diffusion of the C and N interstitial elements but not high enough to permit substitutional diffusion. At these low temperatures, the chromium substitutional element's mobility is sufficiently reduced to inhibit the nucleation and growth of nitrides and/or carbides. Under these conditions, the matrix becomes continuously and increasingly enriched by the interstitial element, promoting a non-equilibrium saturation of the crystalline lattice and stabilizing expanded phases formed in the diffusion layer.

Interstitial supersaturation in the diffusion zone contributes to: (i) formation of interstitially supersaturated phases, (ii) intense interstitial hardening, as a consequence of the colossal amount of interstitial element and its stress fields, contributing to increasing the wear resistance without compromising corrosion resistance [20–24], and (iii) generation of residual compressive stresses in the expanded layer as a result of the restrictive effect of the diffusion-free substrate, which improves fatigue properties [24, 25].

When performing the X-Ray Diffraction of these supersaturated layers, it is observed that the matrix peaks are shifted to lower 2θ angles and show a greater FWHM - full-width at half-maximum height than the peaks of the unenriched matrix phase. This displacement and broadening of the peaks indicate elastic deformation due to the expansion of the crystalline lattice. Zhang and Bell [14] named this phase "Expanded Phase" due to the expansion of the lattice parameters of the crystalline unit cell. On the other hand, Ichii et al. [19] called this phase S-Phase due to the "shifting" to lower angles of the XRD peaks.

Bell and Chen [26] and Sun [27] presented a limit curve for precipitation of chromium nitrides and carbides as a function of temperature and time of plasma diffusion process for an austenitic AISI 316 L stainless steel. "Nitrogen Expanded Austenite (γ_N)" and "Carbon Expanded Austenite (γ_C)" are formed during nitriding or carburizing for temperatures and times below the limit curves shown in **Figure 7**.

Expanded phases obtained at diffusion temperatures between 350°C and 430°C are responsible for surface hardening [20, 21, 28–31]. This hardening can be obtained in all stainless steel families with the formation of different phases expanded by nitrogen and/or carbon [20, 32–36]. **Table 1** shows the different expanded phases formed by low-temperature plasma diffusion surface treatment and their hardening characteristics for different families of stainless steels.

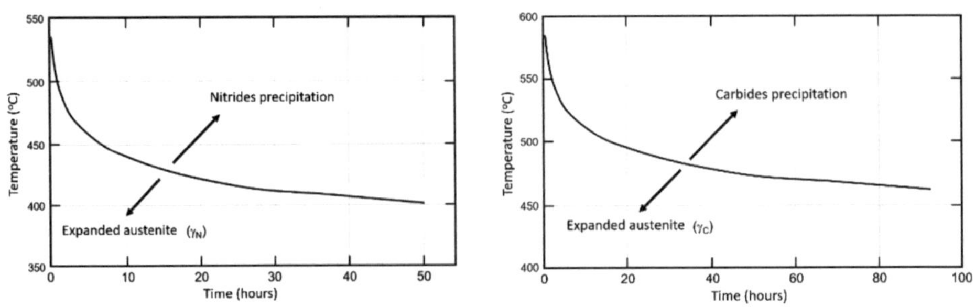

Figure 7.
Limit curves for precipitation of chromium nitrides or chromium carbides in austenite as a function of temperature and time of plasma diffusion process [26, 27].

Steel	Classification	Expanded phase	Symbol	Typical hardness (HV)
Austenitic	AISI 316	Expanded Austenite	γ_N	1400
Martensitic	AISI 420	Expanded Martensite	α'_N	1000
Precipitation-hardening	17-4PH	Expanded Martensite	α'_N	1000
Ferritic	AISI 410S	Expanded Ferrite	α_N	1200
Duplex	AISI F51	Expanded Ferrite	α_N	1000
		Expanded Austenite	γ_N	1000

Table 1.
Expanded phases formed during low-temperature nitriding of stainless steels.

A passivating Cr_2O_3 film formed on the surface of the parts to be nitrided prevents nitrogen or carbon from entering stainless steel. Thus, the passive film's mechanical or chemical removal process should be employed before diffusion. The chemical removal of the passive film by acid pickling may compromise the surface finish of the parts or maybe potential damage to the operators' health or equipment. Currently, modern low-temperature gas nitriding processes still use acid pickling [37] for depassivation of the Cr_2O_3 layer during exposure of the parts' surface to atmospheres containing halides (NF_3 or HCl) has been carried out in Low-temperature gas carburizing [38]. Activation of the parts' surface by nickel plating to prevent repassivation by catalytic decomposition of NH_3 gas [39] has also been used.

Activation of the surface by "sputtering" in H_2, under high voltage and low pressure—[40], use not only the kinetic energy of the ions but also the reducing character of hydrogen [41], preserving the surface quality of the parts being nitrided.

3.1 Austenitic stainless steels

The behavior of nitriding at high temperatures, above 500°C, and at low temperatures, below 420°C, mainly affects the corrosion resistance of the nitrided surface [14]. Austenitic stainless steels cannot be nitrided conventionally at temperatures close to 500–550°C due to intense precipitation of CrN and Cr_2N chromium nitrides in the diffusion zone [42–45]. The precipitation of these nitrides increases the surface hardness but greatly decreases the corrosion resistance due to chromium removal from the solid solution in the matrix. Nitriding must be carried out below 430°C in order to avoid precipitation of nitrides. In this low-temperature nitriding process, generally between 380°C and 420°C, the diffusion kinetics of the chromium substitutional element is significantly reduced, which inhibits the formation of chromium nitrides. The increasing diffusion of nitrogen in the austenite generates a supersaturated solid solution that expands the CFC crystalline lattice and forms the metastable phase called Expanded Austenite—(γ_N) [46–48]. The formation of expanded austenite promotes an increase in surface hardness without compromising corrosion resistance [20, 49, 50].

Figure 8 shows the microstructures obtained by Bruno et al. [50] for AISI 316 L steel after nitriding at temperatures of 550°C (a) and 380°C (b). When this steel is nitrided at 550°C, the nitrided surface becomes dark and severely etched, which denotes the loss of corrosion resistance in this region under the action of Villela's reagent. When nitriding is carried out at 380°C, the nitrided surface appears as a white layer, and the non-nitrided matrix shows a microstructure very similar to a

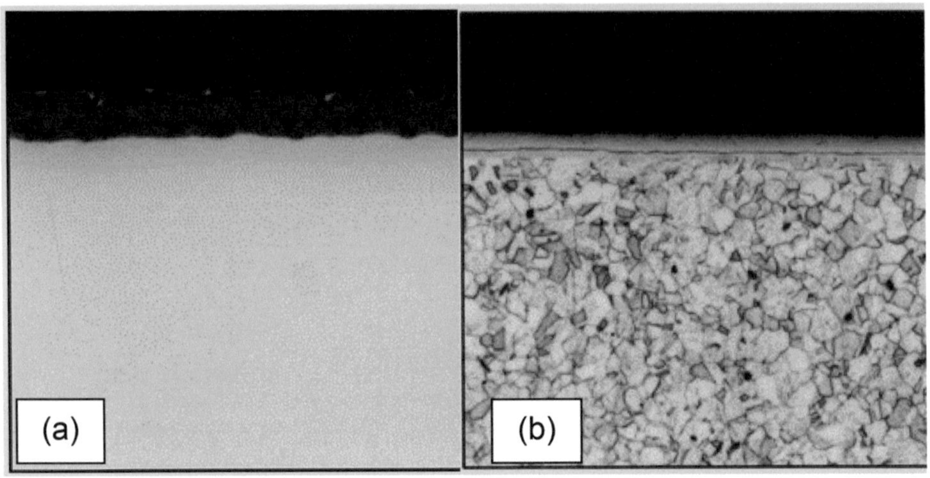

Figure 8.
Images of the nitrided surfaces after nitriding at (a) 380°C and (b) 550°C. Bruno et al. [50].

typical austenitic steel microstructure. In this low-temperature nitriding condition, the Marble metallographic reagent does not etch the nitride layer, only the matrix, which indicates a better corrosion resistance of the nitrided surface. This corrosion resistance behavior against metallographic reagents results from the formation mechanism of expanded phases on the nitrided surface. The X-ray diffractograms in **Figure 9** show the non-nitrided condition after nitriding at 380°C (a) and 550°C (b). In the non-nitrided condition, only FCC austenite peaks are present. When nitriding at 380°C, FCC austenite peaks are shifted to the left and become broader, resulting from residual compression stresses and distortion of the crystalline lattice caused by nitrogen supersaturation. When expanded austenite is formed on the surface of the nitride specimen, the corrosion resistance is maintained or even improved compared to the non-nitrided specimen. On the other hand, when nitriding is carried out at 550°C, several CrN and Cr_2N diffraction peaks show up. Chromium nitride precipitation induces depletion of the Cr content of the metallic matrix and is responsible for the loss of corrosion resistance of the nitrided surface.

The expansion of the FCC crystal lattice and the increase of the lattice parameter, which occurs when expanded austenite is formed, are shown in **Figure 9**. Expanded austenite peaks are shifted to the left, and the volume variation is close to 10% [20]. Strain-free FCC austenite has a lattice parameter equal to 0.359 nm (ICDD® Card 00–033-0397). After plasma nitriding, the lattice parameter in expanded austenite increases to 0.375 nm, corresponding to a calculated nitrogen content at a supersaturation equal to 34.6% atomic or approximately 8.5% by mass. These estimations do not consider the contribution of the residual stresses in shifting the diffraction peaks to the left) [51, 52]. Expanded austenite is responsible for the increased surface hardness up to 7 times over the original hardness, as shown in **Figure 10** [20].

The interstitial supersaturation of the matrix may be due to nitrogen diffusion in nitriding or carbon diffusion [25–54] upon plasma carburizing. The surface treatment may comprise both nitrogen and carbon diffusion, and the plasma treatment is called nitrocarburizing or just carbon for plasma carburizing. These treatments may be carried out at low temperatures, below 430°C for nitrocarburizing and below 500°C for carburizing, avoiding carbide or nitride precipitation. **Figure 11** shows the microstructures after (a) nitriding, consisting of a monolayer of austenite expanded by nitrogen (γ_N); (b) nitrocarburization consisting of a double layer

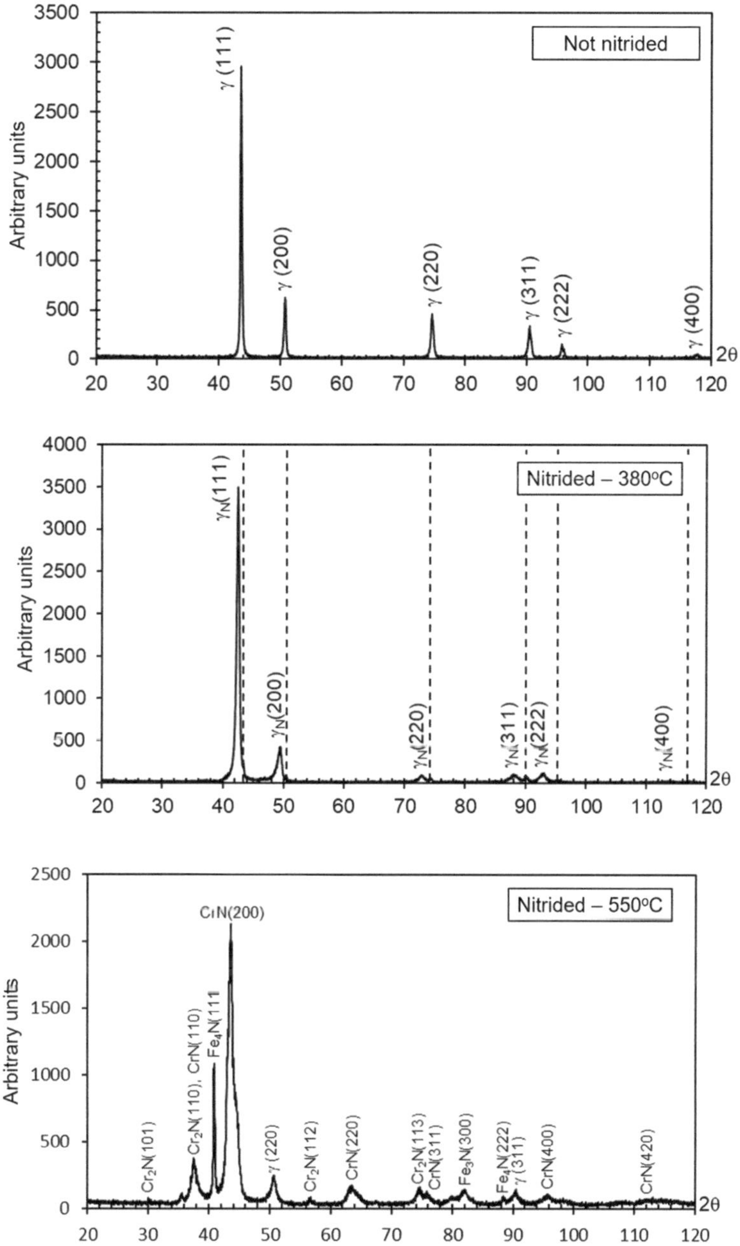

Figure 9.
XRD spectra for AISI 316L steel before and after nitriding [50].

composed of austenite expanded by nitrogen (γ_N) in the outer region and carbon expanded austenite (γ_C) between the first layer and the matrix; and (c) carburizing consisting of a carbon-expanded monolayer of austenite (γ_C).

In nitrocarburizing and carburizing, colossal interstitial supersaturation leads to expansion of the crystalline lattice, generating the expanded phases "γ_N" and "γ_C." **Table 2** shows the expansion characteristics of the austenite FCC crystalline lattice under each condition and the dissolved nitrogen content in the supersaturated condition. The volume expansion of the FCC lattice, indicated by the ratio $\Delta a/a$, is responsible for the hardening and the generation of residual compressive stresses.

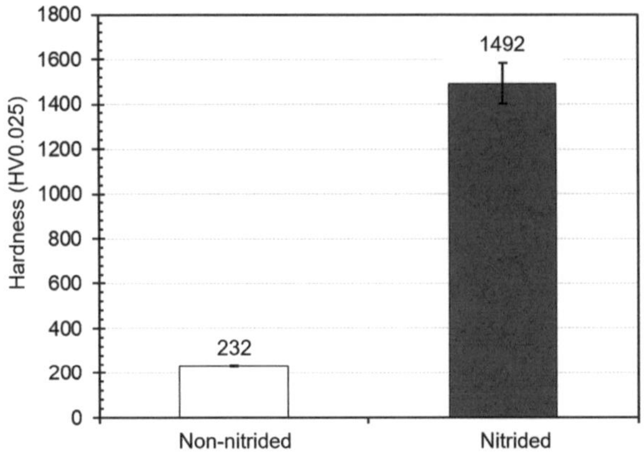

Figure 10.
Surface hardening due to the formation of expanded austenite after plasma nitriding AISI 316L stainless steel at 400°C [20].

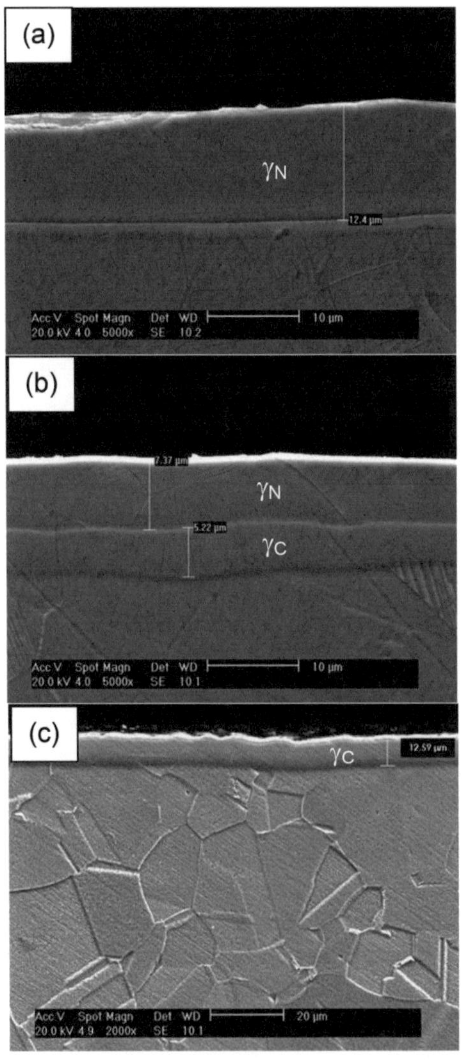

Figure 11.
Microstructures of austenitic stainless steel after (a) nitriding, (b) nitrocarburizing, and (c) carburizing at 400°C. scanning electron microscopy [54].

	Phase	*a* (nm)	$\Delta a/a$ (%)	Concentration, N-C (calculated)
Non-nitrided	γ	0.3589	—	—
Nitrided (400°C)	γ_N	0.3924	9.32	37.2 at.%N
Nitrocarburized	γ_N	0.3916	9.11	36.3 at.%N
(400°C)	γ_C	N.D.	N.D.	N.D.
Carburized (400°C)	γ_C	0.3627	1.05	6.2 at.%C
Carburized (480°C)	γ_C	0.3684	2.64	13.9 at.%C

Table 2.
Lattice parameters, lattice expansion, and calculated interstitial content at supersaturation for the expanded austenite layers in AISI 316L steel.

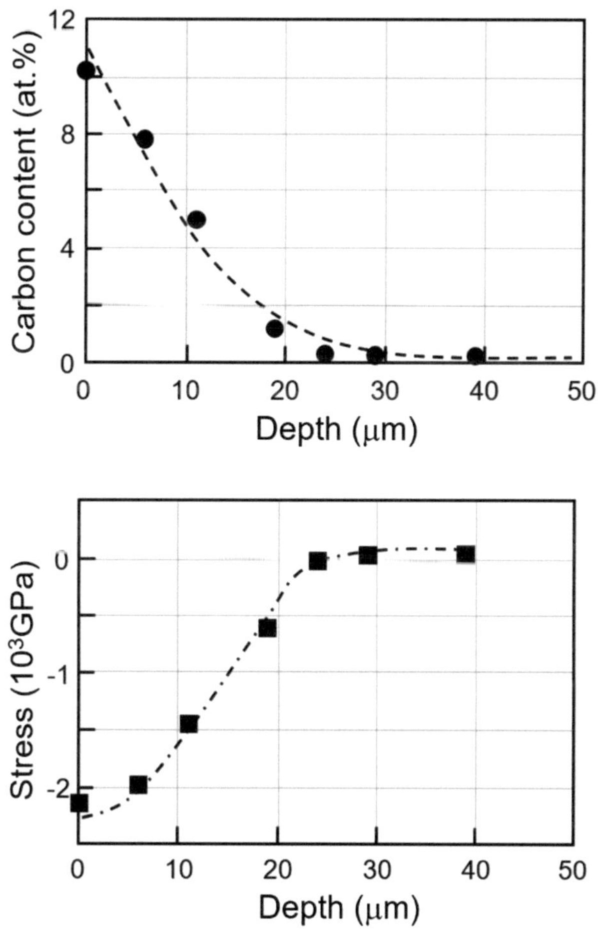

Figure 12.
Carbon pickup and compressive residual stresses on the surface of AISI 316L stainless steel after low temperature carburizing to colossal carbon enrichment [53].

Figure 12 shows the relationship between carbon supersaturation in the FCC crystalline lattice and residual stresses in the case-hardened surface of AISI 316 L austenitic stainless steel. Both parameters gradually decrease toward the nucleus [53] due to supersaturation and the generation of residual compressive stresses on the treated surfaces. **Figure 13** shows the high potential for surface hardening for the three types of treatment [54].

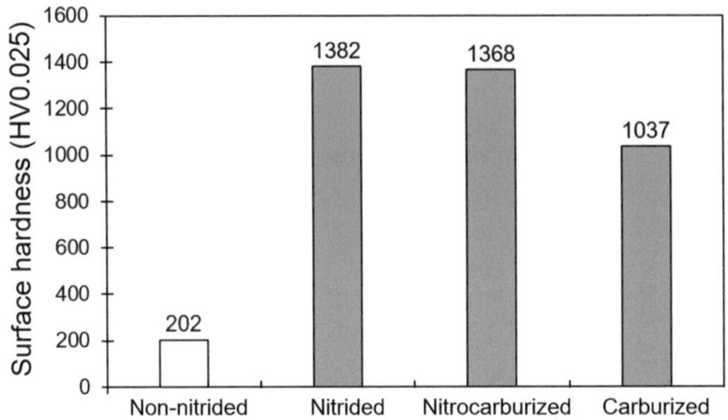

Figure 13.
Maximum hardness on AISI 316L austenitic stainless steel surface upon plasma nitriding, nitrocarburizing, and carburizing at 400°C [54].

3.2 Martensitic stainless steels

Martensitic stainless steels behave similarly to austenitic stainless steels concerning the formation mechanisms of the nitrided surface at different process temperatures. **Figure 14** shows the microstructures of AISI 420 steel after plasma

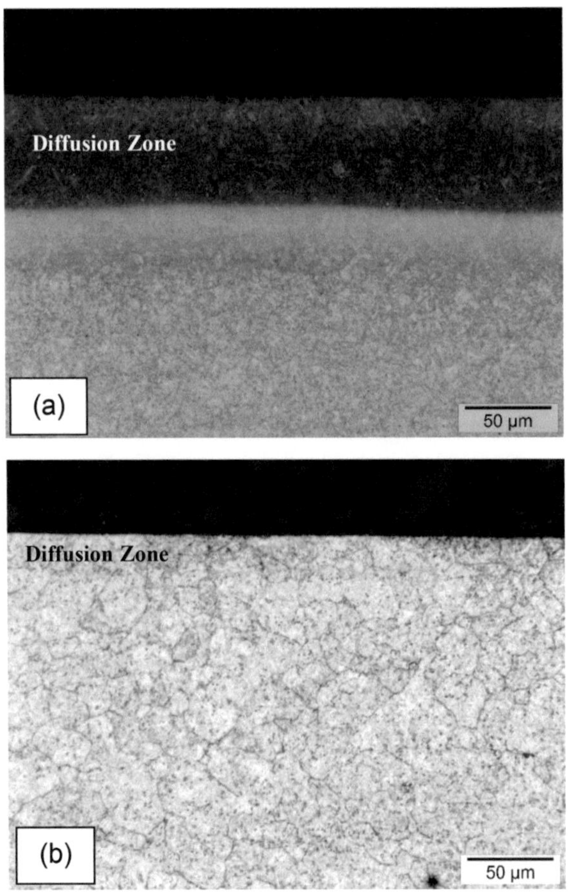

Figure 14.
Microstructures of nitrided surfaces at (a) 550°C and (b) 380°C [33].

nitriding at 380°C and 550°C. At 550°C, the diffusion zone is heavily darkened due to a severe etching by Villela's reagent, but when nitriding is carried out at 380°C, the diffusion zone practically remains unchanged compared to the tempered martensite matrix. The darkening of the nitrided layer denotes loss of corrosion resistance due to the nitriding process, while the unetched nitrided layer indicates that the corrosion resistance is maintained in low-temperature nitriding [33].

In **Figure 15**, the X-ray diffraction maps show AISI 420 steel before and after plasma nitriding [33]. After quenching and tempering, only the peaks referring to the tempered martensite (α′) are observed. When nitriding is carried out at 380°C, the

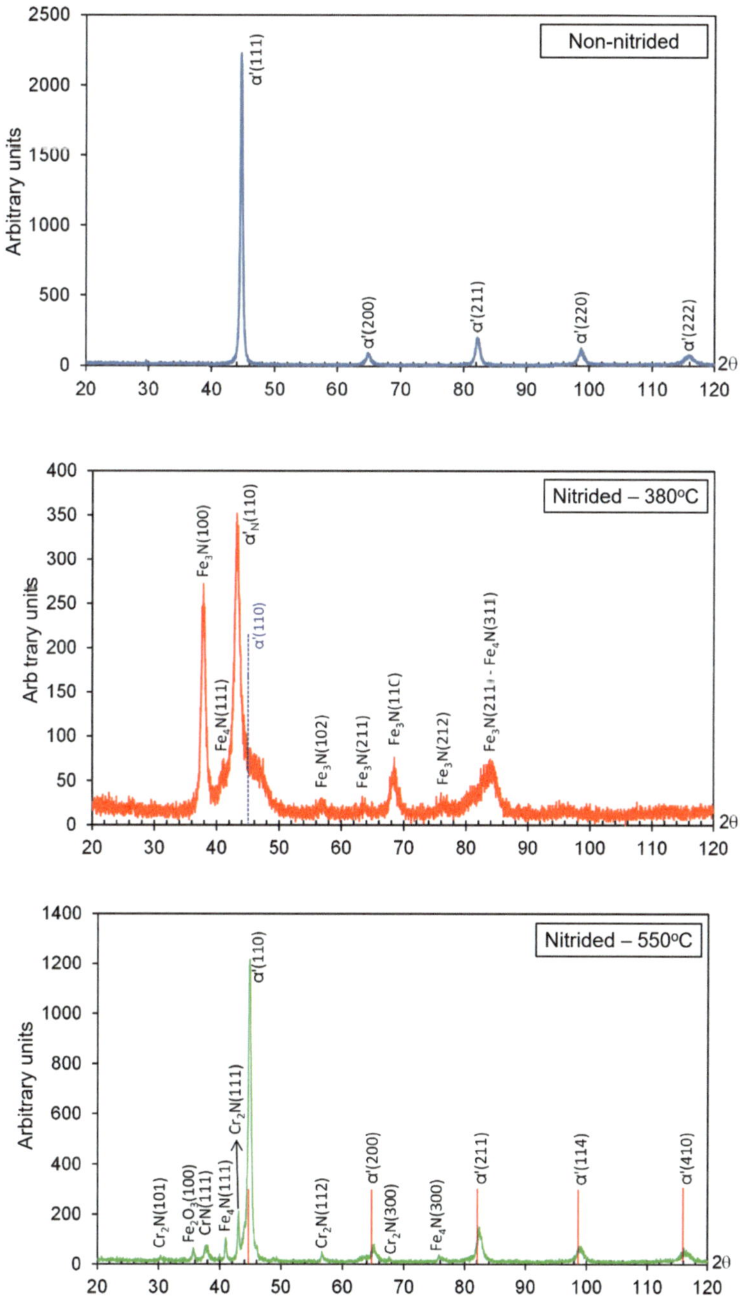

Figure 15.
XRD spectra for AISI 420 steel before and after nitriding [33].

tempered martensite peaks give way to the expanded tempered martensite peak (α'_N). Peaks corresponding to iron nitrides, Fe_3N, and Fe_4N are also observed. This result shows that this temperature is low enough to inhibit the diffusion of chromium, preventing the precipitation of CrN and Cr_2N nitrides. Avoiding the precipitation of chromium nitrides at low temperatures is responsible for maintaining corrosion resistance. The CrN and Cr_2N chromium nitrides diffraction peaks that appear after nitriding at 550°C show intense precipitation of chromium compound and chromium depletion of the matrix, responsible for the decrease in the corrosion resistance of the nitrided surface.

Figure 16 shows the corrosion rate of a 380°C nitrided AISI 420 steel specimen when subjected to an immersion test in an aqueous solution with 10% HCl for 120 h. In the quenched and tempered condition, the corrosion rate after nitriding is lower than the non-nitrided material due to the higher nitrogen concentration on the nitrided surface.

While hardening occurs due to the precipitation of chromium nitrides, in the 550°C plasma nitriding treatment and in the 380°C nitriding treatment, the nitrided surface hardens due to the formation of expanded tempered martensite ($\gamma'N$), which induces compressive residual stresses.

Figure 17 shows that, compared to the quenched and tempered matrix, with 590 HV, the low-temperature nitriding plasma treatment (380°C) promotes hardening near 1000 HV. For the 550°C nitriding, the hardening nearly reaches

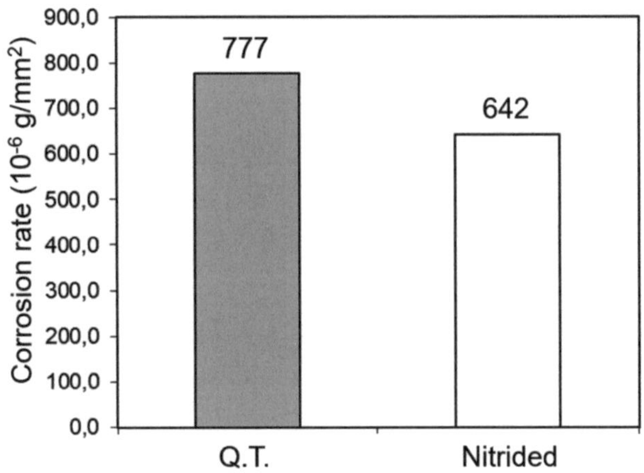

Figure 16.
The corrosion rate of AISI 420 steel in aqueous solution with 10% HCl for 120 h before and after plasma nitriding at 380°C for 20 h. Author: Unpublished.

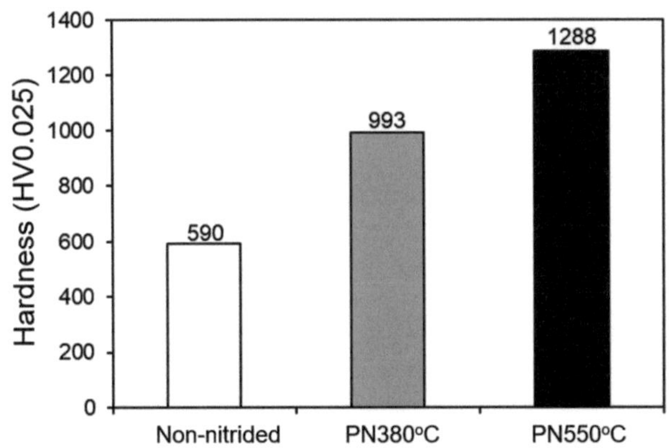

Figure 17.
Maximum hardness after plasma nitriding of AISI 420 steel [32].

1300 HV. Despite the lower hardening in the nitriding treatment at 380°C, this condition should be preferentially used, as it combines hardening and good corrosion resistance.

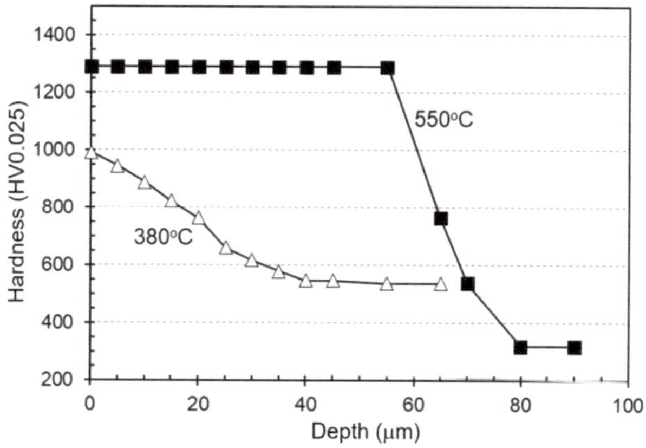

Figure 18.
Transverse hardening profiles after plasma nitriding of AISI 420 steel at 380°C and 550°C [32].

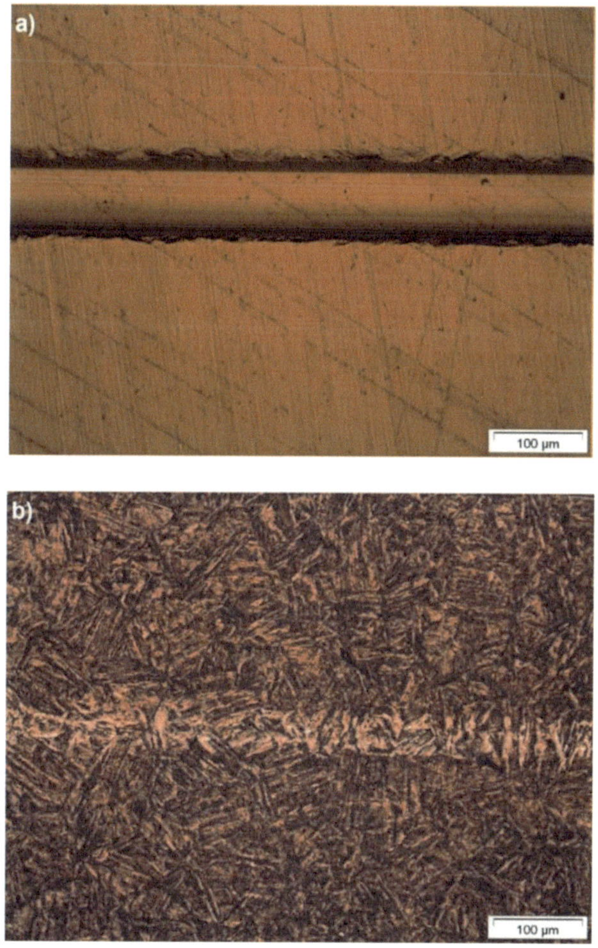

Figure 19.
Scratches made under constant load on the surface of an AISI 410 steel in the (a) quenched and tempered and (b) after plasma nitriding at 400°C conditions [57].

Another important factor related to the hardening characteristic is the transverse hardness profile obtained in these two conditions, **Figure 18**. For the 550°C nitriding treatment, the transverse hardening profile shows a maximum hardness level throughout the diffusion zone with an abrupt drop at the matrix interface [32, 55, 56]. A very steep hardness gradient is not appropriate to withstand mechanical shear stresses found during sliding. Furthermore, exposing the steel to high nitriding temperatures causes a decrease in core hardness by an over-tempering effect [32]. When low-temperature nitriding is carried out, despite the lower maximum hardness, the transverse hardening profile is diffuse, with no decrease in core hardness, and suitable for most different applications.

The surface hardening promoted in the low-temperature plasma nitriding treatment is responsible for increasing the tribological properties [57]. **Figure 19** compares the scratch resistance of an AISI 410 martensitic stainless steel: (a) non-nitrided, quenched, and tempered to a 40 HRC hardness; (b) plasma nitrided at 400°C. The scratch path in the non-nitrided condition is thicker and more profound than in the nitrided condition and presents deformation in its surroundings. **Table 3** shows that the scratch severity is at least half of the non-nitrided condition for the scratch track's depth and thickness in the nitrided condition.

Figure 20 compares the cavitation resistance of non-nitrided and 400° plasma nitrided AISI 410 stainless steel in a test [58]. One can see that the mass loss of the low-temperature plasma nitrided specimens lost 40 times less mass than the non-nitrided specimen.

Martensitic stainless steels can also be nitrocarburized or carburized [59–62]. Nitrocarburization of 420 martensitic stainless steel carried out at 450°C for 4 h can achieve a surface hardening close to 1280 HV with a layer composed of nitrogen and

Condition	Hardness (HV0.01)	Width (μm)	Depth (μm)
Non-nitrided	388	90	26
Nitrided	1275	43	12

Table 3.
Scratch width and depth for non-nitrided and 400°C plasma nitrided AISI 410 stainless steel.

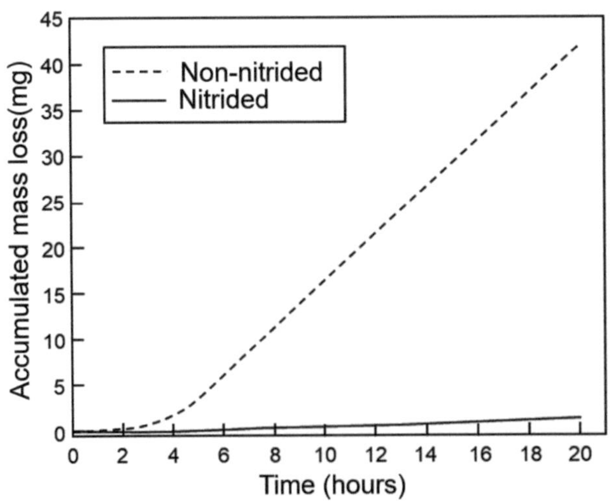

Figure 20.
Mass loss during cavitation tests of an AISI 410 steel in the quenched and tempered and 400°C plasma nitrided conditions [58].

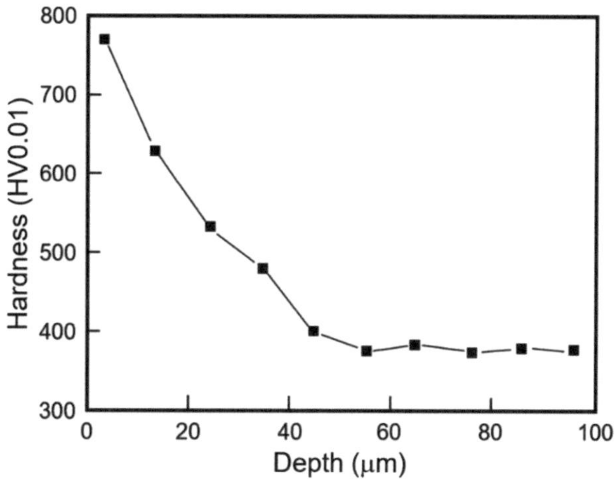

Figure 21.
Hardness profile for a 450°C (4 h) plasma nitrided AISI 420 martensitic stainless steel [61].

carbon expanded martensite (γ'_{NC}) and $Fe_3C/Fe_{2-3}(CN)$ type precipitates. Nitriding at lower temperatures avoids these precipitates in the layer. **Figure 21** shows the hardness profile of the martensitic stainless steel after plasma hardening at 450°C for 4 h, with a maximum hardening potential of 800 HV and a hardening depth in the diffusion zone close to 0.040 mm [61].

3.3 Precipitation hardening stainless steels

The plasma nitriding process for PH precipitation-hardening stainless steels should preferably be carried out at temperatures equal to or below the aging temperature of the steel part. PH steels are aged at different temperatures, specified according to the final desired mechanical properties. A proper selection of the nitriding temperature allows for reaching the desired surface hardness without compromising the quenched pus tempered hardness achieved during aging. The plasma nitriding treatment of PH steel components may be carried out at lower or higher temperatures depending on the application and the operating conditions.

Figure 22(a) shows the nitrided layer and the resulting hardening of aged 17–4PH steel after 4 h at 550°C plasma nitriding treatment. For this condition, the nitrided layer is composed of a diffusion zone formed by the precipitation of iron and chromium nitrides. The precipitation of these nitrides promotes an intense surface hardening, capable of raising the surface hardness to values close to 1300 Vickers [34]. **Figure 22(b)** shows that the 35.3 HRC hardness of the substrate, previously aged for 4 h at 552°C, condition H1025 (AMS 5643 2013), practically remained unchanged after nitriding at 550°C/4 h with a measured value of 34.4 HRC [34]. **Figure 23** shows a gentle hardness profile and a nitriding depth close to 0.65 mm.

Nitriding of PH steels can also be performed at lower temperatures so as not to affect corrosion resistance [63]. **Figure 24** shows the nitrided layer of 17–4PH steel after plasma nitriding at 400°C using the active screen technique. It is observed that the nitrided layer is white and not etched by Villela's reagent, unlike the nitrided layer at 550°C, which is dark and severely etched by Villela's reagent. This difference in behavior is related to the nitriding mechanisms. As hardening in nitriding at 550°C occurs with the precipitation of chromium nitrides, corrosion resistance decreases as the matrix is depleted in chromium. When plasma nitriding is carried out at 400°C, hardening occurs by forming a nitrogen supersaturated layer of expanded martensite (α'_N) without nitrides precipitation, reaching 1130 Vickers.

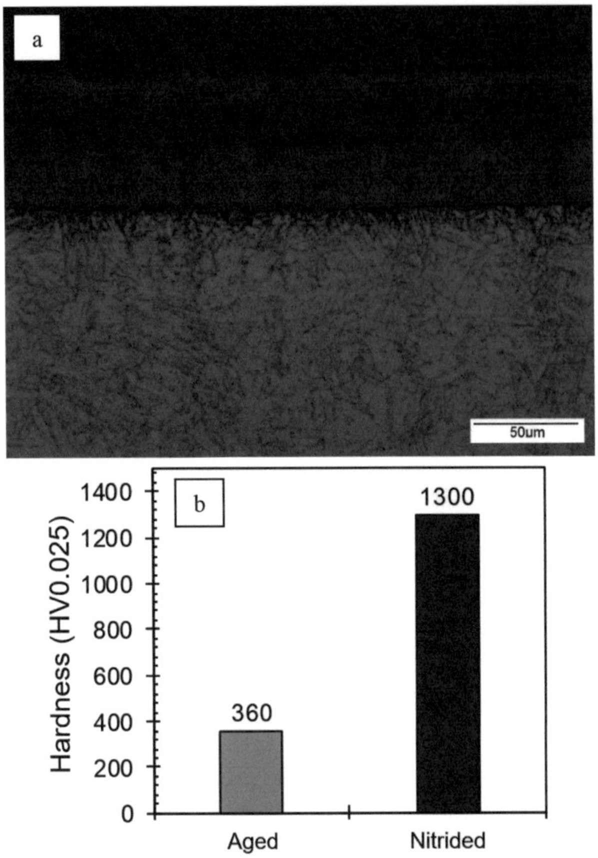

Figure 22.
Nitrided layer (a) and hardening characteristic (b) of 17–4PH steel after plasma nitriding in DC-plasma at 550°C [34].

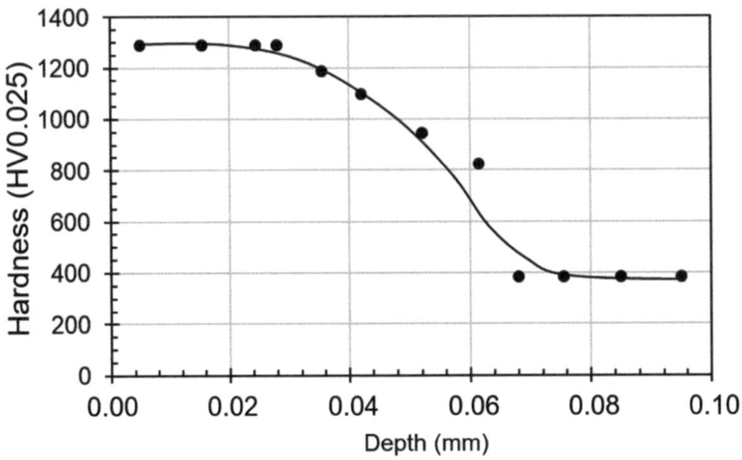

Figure 23.
Transverse hardening profile of 17–4PH steel after DC plasma nitriding at 550°C [34].

3.4 Ferritic and duplex stainless steels

Low-temperature plasma nitriding of ferritic and duplex stainless steels is still being developed and is not commercially available yet. However, many reports

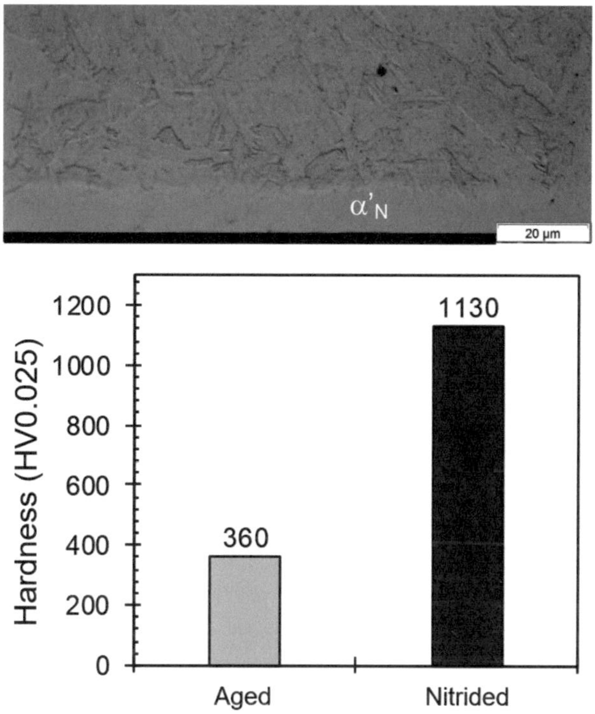

Figure 24.
Hardness variation after plasma nitriding of 17–4PH stainless steel α′N expanded martensite layer and Vickers hardness before and after ASPN at 400°C [63].

promise good results for use in most different components and applications. When ferritic stainless steels are nitrided at low temperatures, precipitation of chromium nitrides is avoided [35, 64]. **Figure 25** shows the microstructure of a plasma nitrided AISI 410S stainless steel with a layer of expanded ferrite (α_N) containing Fe_3N iron nitrides. Shifted to the left, expanded ferrite (α_N) peaks appear on the X-ray

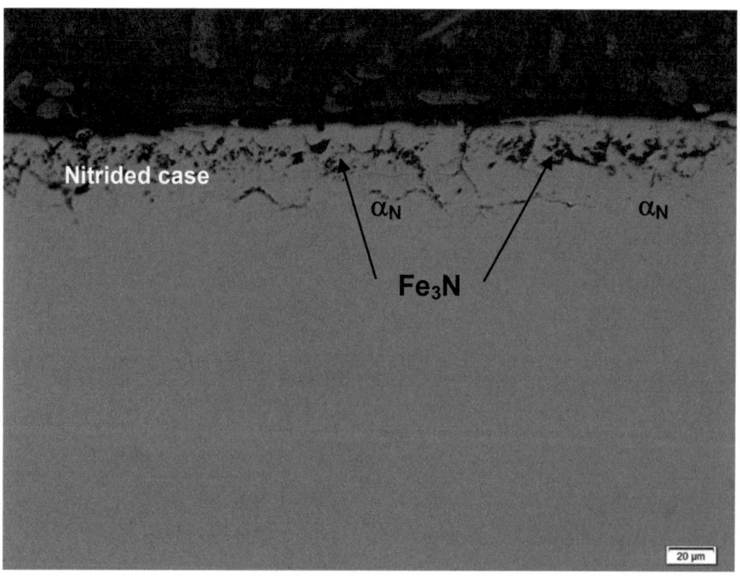

Figure 25.
Microstructure of the nitrided surface after plasma nitriding AISI 410S stainless steel at 400°C [64].

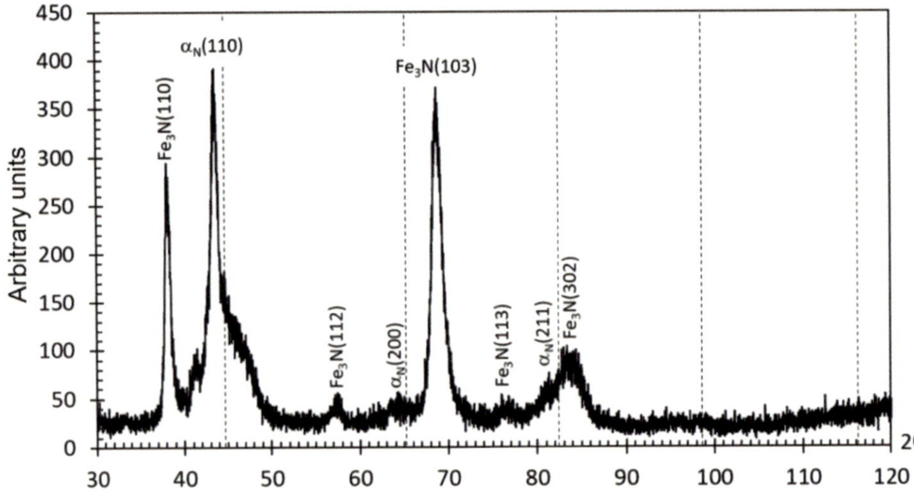

Figure 26.
XR diffraction pattern of the surface after plasma nitriding AISI 410S stainless steel at 400°C [64].

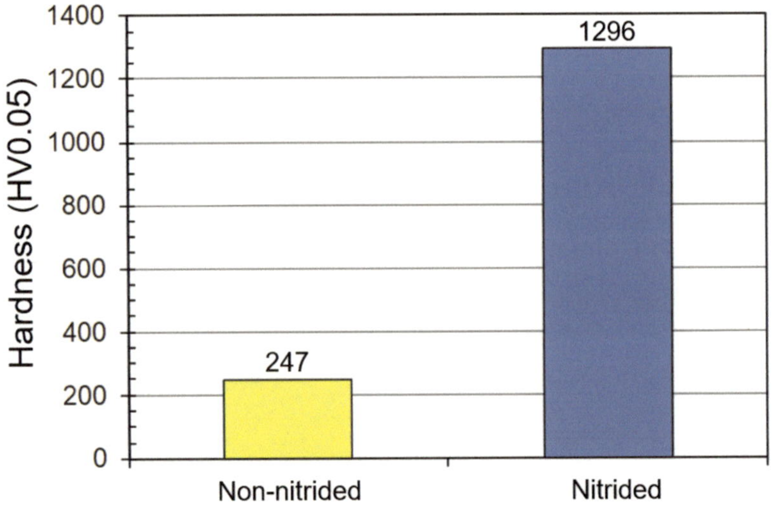

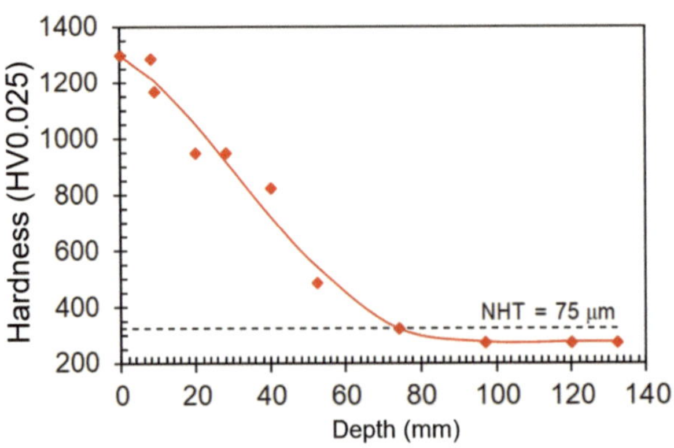

Figure 27.
Surface hardness and hardness profile of an AISI 410S stainless steel after plasma nitriding at 400°C [64].

diffraction pattern of the nitrided layer, **Figure 26**. Vertical dashed lines indicate the positions of ferrite peaks in the matrix. Besides, Fe_3N peaks were also detected. The absence of chromium nitrides in the nitrided layer grants the corrosion resistance of the nitrided surface. **Figure 27** shows the hardening obtained in the nitriding by comparing the maximum surface hardness obtained and the transversal hardening profile of the nitrided surface.

Corrosion resistance testing carried out by immersion in 3% $FeCl_3$ aqueous solution, for 88 h, at room temperature showed a better performance of the nitrided specimens concerning the non-nitrided ones, **Figure 28**. When the steel is nitrided at a low temperature (N400°C), the corrosion properties are not changed compared to the non-nitrided condition. However, high-temperature nitriding (N530°C)

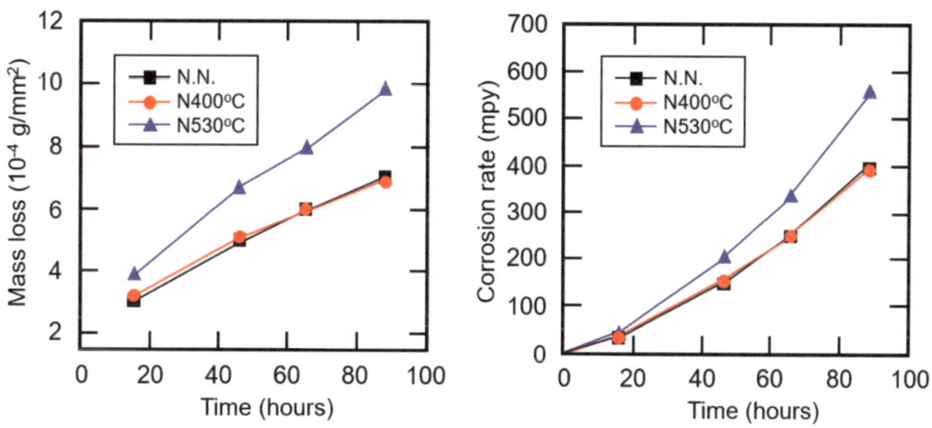

Figure 28.
Mass loss and corrosion rates of non-nitrided and plasma nitrided AISI 410S stainless steel during immersion in 3% $FeCl_3$ aqueous solution for 88 h at room temperature [65].

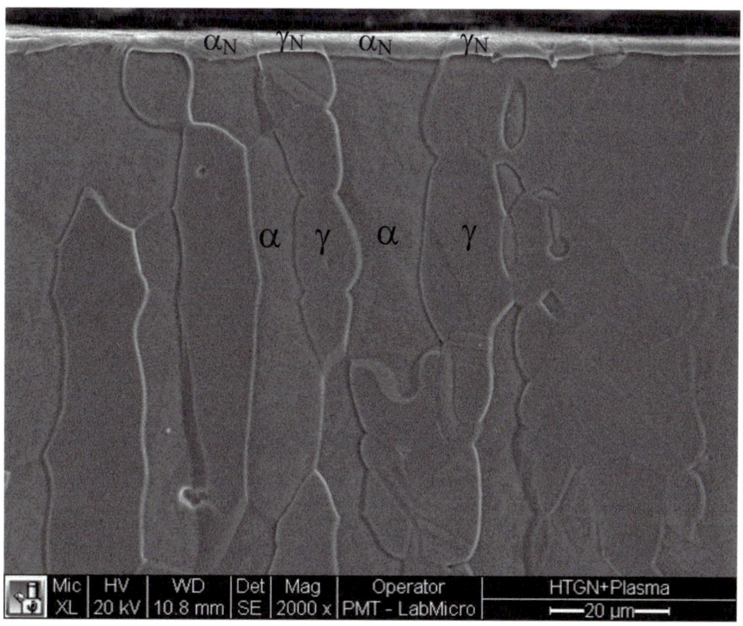

Figure 29.
Microstructure of 2205 duplex stainless steel, after low-temperature plasma nitriding. Expanded ferrite and expanded austenite [67].

promotes a significant loss of corrosion resistance compared to the other two conditions [65].

Duplex stainless steels' microstructure is composed of austenite and ferrite in approximately equal proportions. In this condition, low-temperature nitriding leads to the formation of expanded austenite (γ_N) and expanded ferrite (α_N) on top of ferrite and austenite strings, respectively [36, 66, 67]. **Figure 29** shows the microstructure on the surface of type 2205 duplex stainless steel after plasma nitriding at 400°C. The austenite and ferrite bands and the formation of the respective expanded phases on the nitrided surface are observed in the photomicrograph [67]. The X-ray diffraction pattern in **Figure 30** shows the initial phases' peaks and shifted to the left, the respective peaks of the nitrogen-expanded phases. Fe_3N iron nitrides were also detected [67]. Consequently, the formation of expanded austenite and expanded ferrite on the surface led to an intense hardening of the nitrided surface, as shown in **Figure 31**.

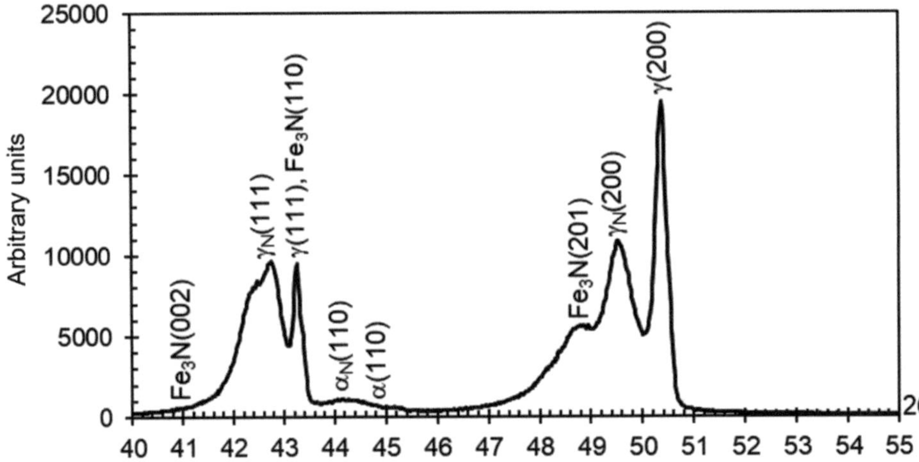

Figure 30.
XR-diffraction pattern of 400°C plasma nitrided 2205 duplex stainless steel [67].

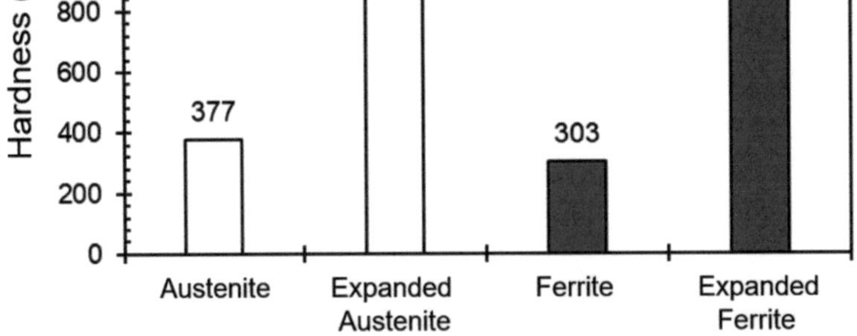

Figure 31.
Hardness of ferrite and austenite and expanded ferrite and expanded austenite in 400°C plasma nitrided 2205 duplex stainless steel [36].

4. Duplex diffusion surface treatments

Duplex stainless steels can also be subjected to a duplex nitriding treatment consisting of a combination of high-temperature and low-temperature diffusion treatments. Firstly, an HTGN—high-temperature gas nitriding is carried out at 1200°C, followed by a 400°C LTPN—low-temperature plasma nitriding aiming for a better load-bearing capacity. In the high-temperature nitriding treatment, nitrogen is introduced on the surface of the steel, shifting the phase equilibrium so that ferrite stringers are transformed to austenite, thus forming a 100 μm thick fully γ

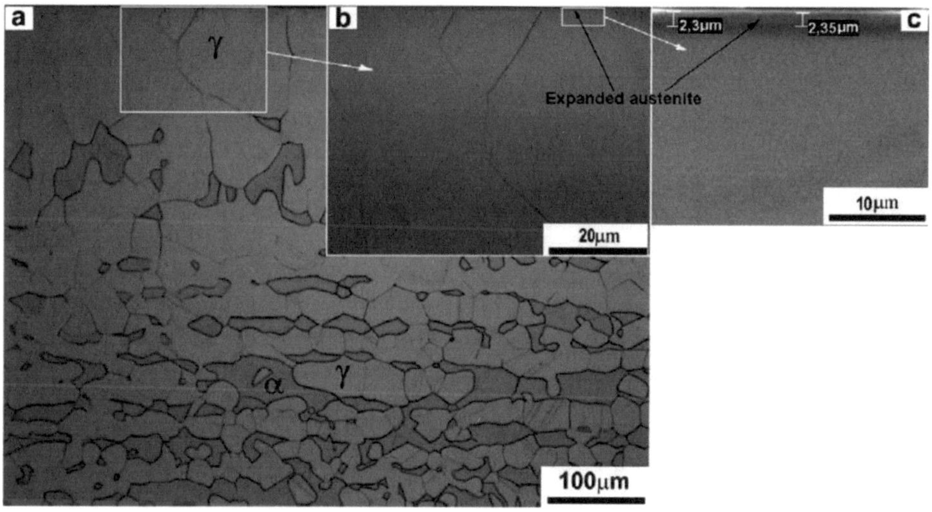

Figure 32.
Microstructure of duplex treated 2205 duplex stainless steel (HTGN + LTPN) [11].

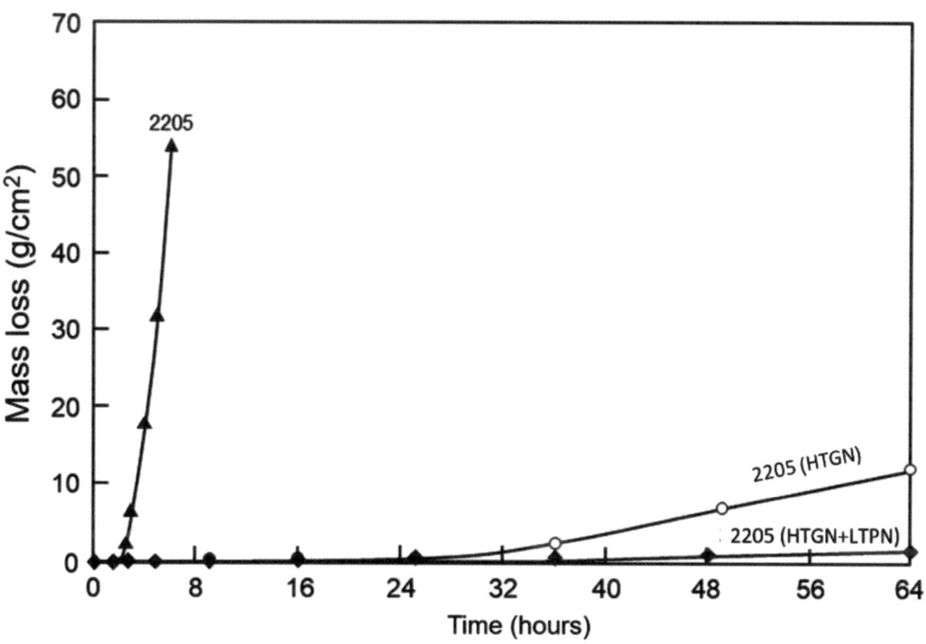

Figure 33.
Mass loss during a cavitation-erosion test. 2205 duplex stainless steel non-nitrided, HTGN, and duplex nitrided (HTGN + LTPN) [11].

layer, raising the hardness from 280 to 330 HV. Subsequently, the LTPN—low-temperature plasma nitriding diffusion treatment leads to a continuous and homogeneous layer of expanded austenite (γ_N), 1200 HV hard, on top of the austenite layer [11]. **Figure 32** shows the microstructure of the duplex nitrided 2205 steel. This microstructure grants greater load-bearing capacity than the single plasma nitriding treatment, and the alloy's performance under cavitation-erosion is much better, as shown in **Figure 33**.

Duplex treatments (HTGN + LTPN) can considerably increase the tribological properties of the surface. **Figure 33** shows the mass loss results during a 2205 duplex-stainless-steel cavitation-erosion test. The mass loss decreases after (HTGN) high-temperature gas nitriding, forming a thick high-nitrogen austenite layer. In the duplex treatment, the mass loss is almost null for testing times up to 64 h due to forming a 1300 HV hard expanded austenite layer sustained by a harder substrate [11].

5. Plasma diffusion surface treatment

A plasma diffusion technology, which combines a plasma nitriding treatment with a subsequent solubilization treatment, was proposed in pioneering work by Pinedo et al. for solid-state alloying [68, 69]. A 1 mm thick sample of AISI 316 L austenitic stainless steel, plasma nitrided at 470°C for 12 h, in a $1N_2$:$1H_2$ gas mixture, formed a 60 μm deep, 1290 HV hard nitrided layer composed of γ + CrN + Cr_2N. After nitriding, the material was solubilized at 1150°C and cooled in water to promote the diffusion and homogeneous redistribution of nitrogen through the sheet's cross-section. After solubilizing, a 0.80 wt.%N, homogeneous nitrogen content was found throughout the thickness, consistent with Thermocalc© predictions. **Figure 34** shows the nitrogen enrichment profile on the nitrided surface, reaching a maximum content of 10 wt.%. **Figure 35** shows the hardness profile along the sheet thickness showing a homogeneous increase in hardness from 200 to 300 HV, achieving complete hardening of the cross-section.

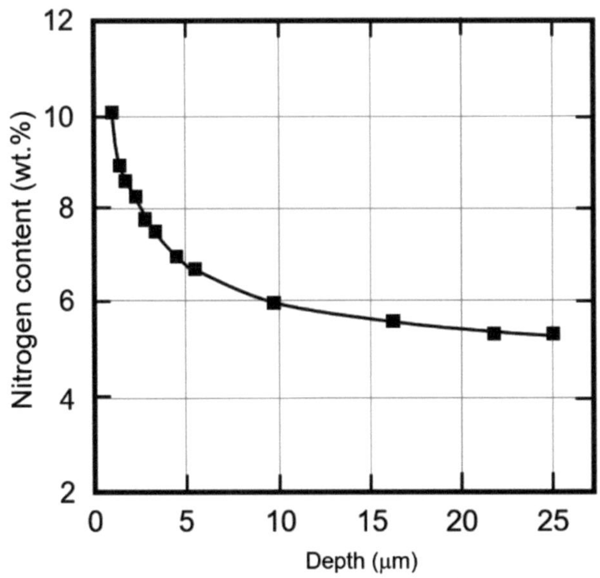

Figure 34.
Compositional profile of nitrogen obtained by GDOES after plasma nitriding of AISI 316 L steel [69].

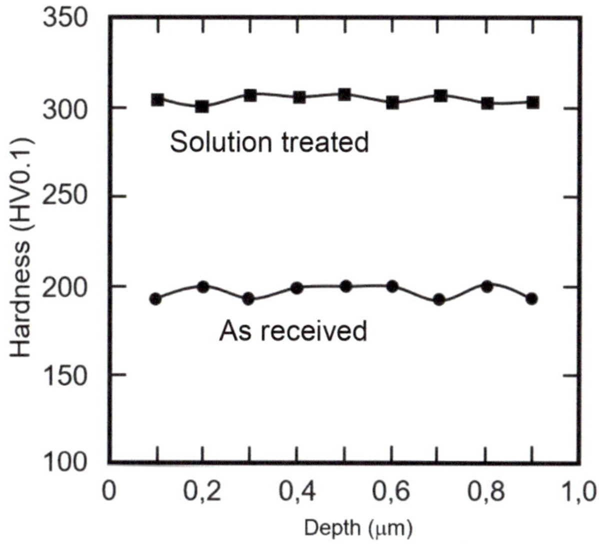

Figure 35.
Transverse hardening of AISI 316 steel nitrided under plasma and solubilized at 1150°C [69].

Recently [70] Berton et al. applied the same principle for an AISI 409 ferritic stainless steel. AISI 409 specimens $0.8N_2$:$0.2H_2$ plasma nitrided at 510°C for 2 h and later solubilized at 1100°C for 1 h for nitrogen diffusion in the ferritic matrix. Once the steel was solubilized, it was subjected to quenching and tempering treatments to obtain a high surface hardness tempered martensite layer. **Figure 36** presents the nitrogen profile along the cross-section for the solubilized (NS) and tempered conditions from 950°C (Q950) and 1050°C (Q1050), showing a maximum effective enrichment of the order of 1% by mass. This enrichment after the complete quenching and tempering cycles promotes a maximum rise in surface hardness close to 4x the core hardness at transverse hardening depths of up to 600 μm, as shown in **Figure 37**.

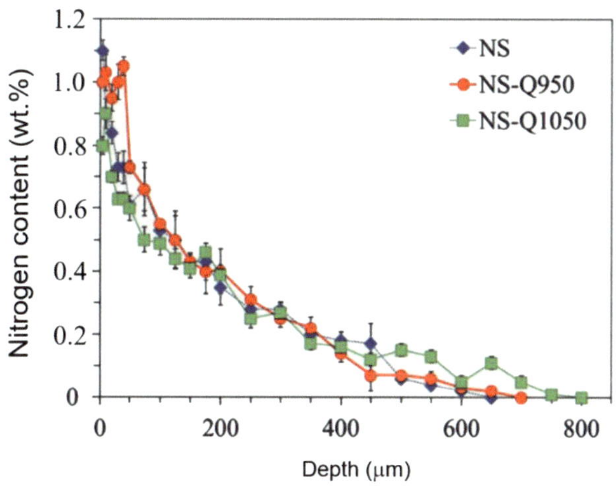

Figure 36.
Compositional profile of nitrogen determined by WDS in the diffused condition (NS) and after quenching from 950°C (NS-Q950) and 1050°C (NSQ1050) [70].

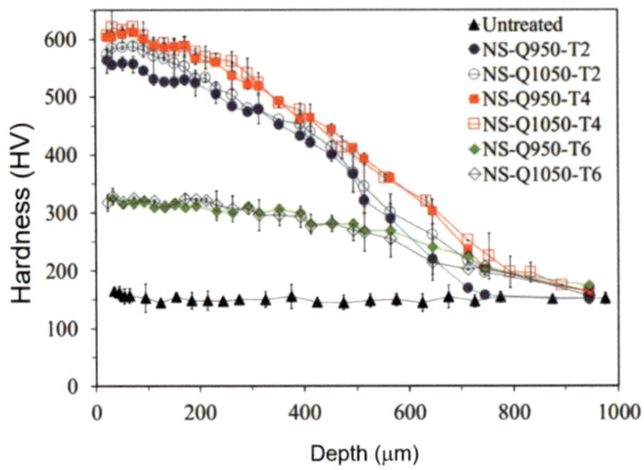

Figure 37.
Transverse hardening profiles of AISI 409 steel under conditions; solubilized (NS) and after quenching at 950°C (Q950) and 1050°C (Q1050), and tempering for 1 h at 250 (T2), 450 (T4), and 650°C (T6). The untreated condition is shown for comparison [70].

6. Conclusion

Although stainless steels are designed to sustain distinctly superior corrosion resistance in a wide range of aggressive environments, these steels do not show enough wear resistance in many highly demanding tribological conditions, harming performance and service life.

Diffusion-based surface hardening processes are an alternative to increasing the wear resistance of stainless steel without compromising its corrosion properties. Accordingly, different thermochemical solutions of diffusion surface treatment of stainless steels were presented; (i) high-temperature gas nitriding, (ii) low-temperature plasma or gas nitriding, nitrocarburizing, or carburizing, (iii) duplex treatment combining high and low-temperature nitriding treatments, and (iv) solid-state annealing, to promote surface hardening and maintain or even raise, the corrosion resistance of these materials.

Surface hardening by diffusion thermochemical processes is an efficient strategy to produce tailor-made surfaces with improved mechanical strength and wear resistance, applicable to all classes of stainless steel.

Acknowledgements

The authors acknowledge the financial support of the Fundação de Amparo à Pesquisa do Estado de São Paulo, process FAPESP 2019/18572-7.

Author details

André Paulo Tschiptschin[1]* and Carlos Eduardo Pinedo[2]*

1 Metallurgical and Materials Engineering Department, University of São Paulo, São Paulo, Brazil

2 Heat Tech—Heat Treatment and Surface Engineering Ltd, Mogi das Cruzes, Brazil

*Address all correspondence to: antschip@usp.br and pinedo@heattech.com.br

References

[1] Berns H, Siebert S. Formation of high nitrogen case on stainless steels. Harterei-Technische Mitteilungen. 1994;**49**:123-129

[2] Sieverts A. The absorption of gases by metals. Zeitschrift für Metallkunde. 1929;**21**:37-46

[3] Anderson J.O, Helander T, Höglund L. Thermocalc and Dictra, computational tools for materials science. Calphad 2002;**26**:273-312. DOI: 10.1016/S0364-5916(02)00037-8

[4] Garzon CM, Tschiptschin AP. Thermodynamic and kinetic modeling of the thermal and thermochemical processing of steels using the Calphad method. Matéria. 2006;**11**:70-87. DOI: 10.1590/S1517-70762006000200002

[5] Berns H. Solution nitriding of stainless steels. In: Bell T, Akamatsu K, editors. Stainless Steel 2000—Thermochemical Surface Engineering of Stainless Steel. 1st ed. London: CRC Press; 2000. pp. 111-116. DOI: 10.1201/9780367814151

[6] Tschiptschin AP. Predicting microstructure development during high temperature nitriding of martensitic stainless steels using thermodynamic modeling or high-temperature-solution-nitriding. Materials Research. 2002;**5**:257-262. DOI: 10.1590/S1516-14392002000300007

[7] Christiansen TL, Villa M, Tibollo C. High temperature solution nitriding of stainless steels; current status and future trends. HTM Journal of Heat Treatment and Materials. 2022;**2**:69-82. DOI: 10.3139/105.110406

[8] Santos JF, Garzón CM, Tschiptschin AP. Improvement of the cavitation erosion resistance of an AISI 304L austenitic stainless steel by high temperature gas nitriding. Materials Science & Engineering. A, Structural Materials: Properties, Microstructure and Processing. 2004;**382**:378-386. DOI: 10.1016/j.msea.2004.05.003

[9] Lopez DM, Alonso N, Tschiptschin AP. Corrosion-erosion behavior of austenitic and martensitic high nitrogen stainless steels. Wear. 2007;**263**:347-354. DOI: 10.1016/j.wear.2007.01.053

[10] Tschiptschin AP. Behavior of stainless steels exposed to nitrogen atmospheres at high temperatures. [Habilitation thesis]. Printed - Escola Politécnica da Universidade de São Paulo (not online): Polytechnique School, University of São Paulo; 2000

[11] Mesa DH, Pinedo CE, Tschiptschin AP. Improvement of the cavitation erosion resistance of UNS S31803 stainless steel by duplex treatment. Surface & Coatings Technology. 2010;**2010**(205):1552-1556. DOI: 10.1016/j.surfcoat.2010.10.014

[12] Garzón CM, Tschiptschin AP. Nitretação gasosa em alta temperatura de aços inoxidáveis. Matéria. 2005;**10**:502-525

[13] Tschiptschin AP. Powder metallurgy aspects of high nitrogen stainless steels. In: Kamachi Mudali K, Raj B, editors. High Nitrogen Steels and Stainless Steels: Manufacturing, Properties and Applications. 1st ed. ASM International; Woodhead Publishing; 2004. pp. 94-111

[14] Zhang ZL, Bell T. Structure and corrosion resistance of plasma nitrided stainless steel. Surface Engineering. 1985;**1**:131-136. DOI: 10.1179/sur.1985.1.2.131

[15] Stagno E, Pinasco MR, Palombarini G. Behaviour of sintered

410 low carbon steels towards ion nitriding. Journal of Alloys and Compounds. 1997;**247**:172-179. DOI: 10.1016/S0925-8388(96)02604-7

[16] Marchev K, Cooper CV, Giessen BC. Observation of a compound layer with very low friction coefficient in lon-nitriding martensitic 410 stainless steel. Surface and Coatings Technology. 1998; **99**:229-233. DOI: 10.1016/S0257-8972 (97)00533-1

[17] Pinedo CE. Estudo morfológico e cinético da nitretação por plasma do aço inoxidável martensítico AISI 420 [thesis]. Brazil: University of São Paulo; 1990

[18] Pinedo CE, Monteiro WA. Characterization of plasma nitrided case on martensitic stainless steel by scanning electron microscopy. Acta Microscópica. 2001;**1**:315-316

[19] Ichii K, Fujimura K, Takase T. Technol. Rep. Kansai Univ. 1986;27: 135-144, apud Ichii K, Fujimura K, Takase T.Structure of the ion-Nitrided layer of 18-8 stainless steel. In: Bell T, Akamatsu K, editors. Stainless Steel 2000—Thermochemical Surface Engineering of Stainless Steel. 1st ed. London: CRC Press; 2000. pp. 13-22. DOI: 10.1201/9780367814151

[20] Mingolo N, Tschiptschin AP, Pinedo CE. On the formation of expanded austenite during plasma nitriding of an AISI 316L austenitic stainless steel. Surface & Coatings Technology. 2006;**201**:4215-4218. DOI: 10.1016/j.surfcoat.2006.08.060

[21] Dong H. S-phase surface engineering of Fe-Cr, Co-Cr and Ni-Cr alloys. International Materials Reviews. 2010;**55**:65-98. DOI: 10.1179/095066009X12572530170589

[22] Tromas C, Stinville JC, Templier C. Hardness and elastic modulus gradients in plasma-nitrided 316L polycrystalline stainless steel investigated by nanoindentation tomography. Acta Materialia. 2012;**60**:1965-1973. DOI: 10.1016/j.actamat.2011.12.012

[23] Casteletti LC, Lombardi Neto A, Totten GE. Nitriding of stainless steels. Metallography, Microstructure, and Analysis. 2014;**3**:477-508. DOI: 10.1007/s13632-014-0170-4

[24] Somers MAJ, Christiansen TL, Winther G. Expanded austenite; from fundamental understanding to predicting composition- and stress-depth profiles. In: Proceedings of the European Conference on Heat Treatment; Nitriding and Nitrocarburising (ECHT'18); 12–13 April 2018. Friedrichshafen-Germany, Bremen: Stiftung Institut für Werkstofftechnik; 2018. pp. 92-101. Accessed at the DTU Orbit site: https://orbit.dtu.dk/en/publications/expanded-austenite-from-fundamental-understanding-to-predicting-c

[25] Collins SR, Williams PC, Marx SV. Low-Temperature Carburization of Austenitic Stainless Steels. ASM Handbook Heat Treating of Irons and Steels. Vol. 4D. Ohio: ASM International; 2014. pp. 451-460

[26] Bell T, Li CX. Stainless steel low temperature nitriding and carburizing. Advanced Materials & Processing. 2002;**160**:49-51

[27] Sun Y. Kinetics of low temperature plasma carburizing of austenitic stainless steels. Journal of Materials Processing Technology. 2005;**168**: 189-194. DOI: 10.1016/j.jmatprotec.2004.10.005

[28] Sun Y, Li X, Bell T. Low Temperature plasma carburizing of austenitic stainless steels for improved wear and corrosion resistance. Surface Engineering. 1999;**15**: 49-54. DOI: 10.1179/026708499322911647

[29] Fewell MP, Mitchell DRG, Priest JM. The nature of expanded austenite. Surface and Coatings Technology. 2000; **131**:300-306. DOI: 10.1016/S0257-8972(00)00804-5

[30] Pinedo CE. Nitretação por plasma de aços inoxidáveis. Metalurgia e Materiais. 2002;**60**:162-164

[31] Borgioli F. The "expanded" phases in the low-temperature treated stainless steels: A review. Metals. 2022;**12**:331. DOI: 10.3390/met12020331

[32] Pinedo CE, Monteiro WA. Tratamento térmico e de nitretação sob plasma do aço inoxidável martensítico AISI 420. Tecnologia em Metalurgia, Materiais e Mineração. 2011;**8**:86-90

[33] Pinedo C.E, Magnabosco R. Mecanismos de nitretação sob plasma do aço inoxidável martensítico AISI 420 nitretado a alta e baixa temperatura. Tecnologia em Metalurgia, Materiais e Mineração. 2015;**12**:257-264. DOI: 10.4322/2176-1523.0844

[34] Vitoi AA, Magnabosco R, Pinedo CE. Nitretação sob plasma por tela ativa e plasma direto/DC do aço inoxidável endurecível por precipitação tipo 17-4 PH. In: 70° Congresso Anual da Associação Brasileira de Metalurgia, Materiais e Mineração (ABM'15), 17-21 August 2015. Rio de Janeiro, Brazil; 2015. pp. 1249-1258. DOI: 10.5151/1516-392X-26672

[35] Umemura MT, Pinedo CE, Varela LB. Assessment of tribological properties of plasma nitrided 410S ferritic-martensitic stainless steels. Wear. 2019;**426–427**:49-58. DOI: 10.1016/j.wear.2018.12.092

[36] Pinedo CE, Varela LB, Tschiptschin AP. Low-temperature plasma nitriding of AISI F51 duplex stainless steel. Surface & Coatings Technology. 2013;**232**:839-843. DOI: 10.1016/j.surfcoat.2013.06.109

[37] Expanite [Internet]. 2020. Avaiable from: https://www.expanite.com/implementation-of-expanite-surface-hardening/ [Acessed: 29 September, 2021]

[38] Williams PC, Marx SV. Low temperature case hardening processes. US Patent US 6,461,448 B1. Swagelock Co.; 2002

[39] Somers MAJ, Christiansen T, Møller P. Case hardening of stainless steel. Patent deposited at the INPI - Instituto Nacional de Propriedade Insustrial, the Brazilian Insitute for filing patents. WIPO Patent no. WO20047789; 2004

[40] Tschiptschin AP, Pinedo CE. Processo de enriquecer camada superficial de ligas inoxidáveis sob plasma. Brasil. PI09028684: Patent deposited at the INPI—Instituto Nacional da Propriedade Industrial; 2009 Filed: 11/05/2009; 59 Concession: 21/05/2019

[41] Pinedo CE. Nitretação sob plasma—fundamentos e aplicações. Parte II—Aplicação em aços inoxidáveis. Industrial Heating. 2012;**4**:28-31

[42] Czerwiec T, Renevier N, Michel H. Low-temperature plasma-assisted nitriding. Surface and Coatings Technology. 2000;**131**:267-277. DOI: 10.1016/S0257-8972(00)00792-1

[43] Venkatesan K, Subramanian C, Green LK. Influence of chromium content on corrosion of plasma-nitrided steels. Corrosion. 1997;**53**:507-515. DOI: 10.5006/1.3290282

[44] Liang W, Bin X, Zhiwe Y. The wear and corrosion properties of stainless steel nitrided by low-pressure plasma-arc source ion nitriding at low temperatures. Surface and Coatings

Technology. 2000;**130**:304-308. DOI: 10.1016/S0257-8972(00)00713-1

[45] Liang W, Xiaolei X, Bin X. Low temperature nitriding and carburizing of AISI304 stainless steel by a low pressure plasma arc source. Surface and Coatings Technology. 2000;**1–3**:563-567. DOI: 10.1016/S0257-8972(00)00836-7

[46] Menthe E, Rie K-T. Further investigation of the structure and properties of austenitic stainless steel after plasma nitriding. Surface and Coatings Technology. 1999;**199**:116-119. DOI: 10.1016/S0257-8972(99)00085-7

[47] Pranevicius L, Templier C, Rivière JP. On the mechanism of ion nitriding of an austenitic stainless steel. Surface and Coatings Technology. 2001; **135**(2–3):250-257. DOI: 10.1016/S0257-8972(00)00988-9

[48] Picard S, Memet JB, Sabot R. Corrosion behaviour, microhardness and surface characterization of low energy, high current ion implanted austenitic stainless steel. Materials Science and Engineering A. 2001;**303**: 163-172. DOI: 10.1016/S0921-5093(00)01841-4

[49] Tschiptschin AP, Pinedo CE. Estrutura e propriedades do aço inoxidável austenítico AISI 316L Grau ASTMF138 nitretado sob plasma à baixa temperatura. Revista Escola de Minas. 2010;**63**:137-141

[50] Bruno GT, Boccalini M Jr, Pinedo CE. Tratamento combinado de nitretação sob plasma e revestimento PVD em aço inoxidável austenítico AISI 316L. In: Proceedings of 67° Congresso Anual da Associação Brasileira de Metalurgia, Materiais e Mineração (ABM'12). 31 July–03 August 2012. Rio de Janeiro, Brazil; 2012. pp. 2505-2516

[51] Christiansen TL, Hummelshøj TS, Somers MAJ. Expanded austenite, crystallography and residual stress. Surface Engineering. 2010;**26**:242-247. DOI: 10.1179/026708410X12506870724316

[52] Sato FL, Espitia LA, Pinedo CE. Uso de ensaios de microesclerometria instrumentada no estudo das propriedades da austenita expandida. Tecnologia em metalurgia, materiais e mineração. 2015;**12**:115-122. DOI: 10.4322/2176-1523.0827

[53] Michal G.M, Ernst F, Kahn H. Carbon supersaturation due to paraequilibrium carburization: Stainless steels with greatly improved mechanical properties. Acta Materialia 2006:54: 1597-1606. DOI: 10.1016/j.actamat.2005.11.029

[54] Pinedo CE, Tschiptschin AP. Low temperature nitriding, nitrocarburising and carburising of AISI 316L austenitic stainless steel. International Heat Treatment and Surface Engineering. 2011;**5**:73-77. DOI: 10.1179/174951411X13051201040703

[55] Pinedo CE, Monteiro WA. Surface hardening by plasma nitriding on high chromium alloy steel. Journal of Materials Science Letters. 2001;**20**: 147-149

[56] Pinedo CE, Monteiro WA. On the kinetics of plasma nitriding a martensitic stainless steel type AISI 420. Surface & Coatings Technology. 2004; **179**:119-123. DOI: 10.1016/S0257-8972(03)00853-3

[57] Espitia LA, Dong H, Li X-Y, Pinedo CE. Scratch test of active screen low temperature plasma nitrided AISI 410 martensitic stainless steel. Wear. 2017;**376–377**:30-36. DOI: 10.1016/j.wear.2017.01.091

[58] Espitia LA, Dong H, Li X-Y. Cavitation erosion resistance and wear mechanisms of active screen low

temperature plasma nitrided AISI 410 martensitic stainless steel. Wear. 2015; **332–333**:1070-1079. DOI: 10.1016/j.wear.2014.12.009

[59] Anjos AD, Scheuer CJ, Brunatto SF. Low-temperature plasma nitrocarburizing of the AISI 420 martensitic stainless steel: Microstructure and process kinetics. Surface & Coatings Technology. 2015; **275**:51-57. DOI: 10.1016/j.surfcoat.2015.03.039

[60] Liu R, Yan M. Characteristics of AISI 420 stainless steel modified by low-temperature plasma carburizing with gaseous acetone. Coatings. 2019;**9**:1-10

[61] Scheuer CJ, Rodrigo P, Cardoso RP. Effects of the voltage and pressure on the carburizing of martensitic stainless steel in pulsed DC glow discharge. Materials Research. 2021;**24**:1-11. e20210154. DOI: 10.1590/1980-5373-MR-2021-0154

[62] Scheuer CJ, Cardoso RP, Pereira R. Low temperature plasma carburizing of martensitic stainless steel. Materials Science and Engineering A. 2012;**539**: 369-372. DOI: 10.1016/j.msea.2012.01.085

[63] Pinedo CE, Larrotta SIV, Nishikawa AS. Low temperature active screen plasma nitriding of 17-4PH stainless steel. Surface and Coatings Technology. 2016;**308**:189-194. DOI: 10.1016/j.surfcoat.2016.07.096

[64] Pinedo CE, Magnabosco R, Silva WR. Formação de ferrita expandida na nitretação de um aço inoxidável ferrítico. In: Proceedings of the 72° Congresso Anual da Associação Brasileira de metalurgia, Materiais e Mineração (ABM'17); 2–6 October 2017. São Paulo, Brazil; 2017. pp. 104-114. DOI: 10.5151/1516-392X-30102

[65] Varela LB, Umemura MT, Calderón-Hernández JW. Corrosion resistance of low-temperature and conventional plasma-nitrided 410S ferritic-martensitic stainless steels. Materials Performance and Characterization. 2021;**10**:181-188. DOI: 10.1520/MPC20200099

[66] Pintaude G, Rovani AC, Klein das Neves JC. Wear and corrosion resistances of active screen plasma-nitrided duplex stainless steels. Journal of Materials Engineering and Performance. 2019;**28**:3673-3682. DOI: 10.1007/s11665-019-04114-y

[67] Tschiptschin AP, Varela LB, Pinedo CE. Development and microstructure characterization of single and duplex nitriding of UNS S31803 duplex stainless steel. Surface & Coatings Technology. 2017;**327**:83-92. DOI: 10.1016/j.surfcoat.2017.08.018

[68] Graviljuk V, Berns H. High Nitrogen Steels—Structure, Properties, Manufacture, Applications. Springer-Verlag Berlin Heidelberg GmbH. p. 218. DOI: 10.1007/978-3-662-03760-7

[69] Pinedo CE, Vatavuk J, Oliveira SD. Solid state alloying by plasma nitriding and diffusion annealing for austenitic stainless steel. Material Science Forum. 1999;**318-320**:233-240. DOI: 10.4028/www.scientific.net/MSF.318-320.233

[70] Berton EM, Klein das Neves JC, Mafra M. Quenching and tempering effect on the corrosion resistance of nitrogen martensitic layer produced by SHTPN on AISI 409 steel. Surface & Coatings Technology. 2020;**395**:125921. DOI: 10.1016/j.surfcoat.2020.125921

Chapter 6

Surface Integrity of Ball Burnished 316L Stainless Steel

Selma Attabi, Abdelaziz Himour, Lakhdar Laouar and Amir Motallebzadeh

Abstract

316L is a type of austenitic stainless steel that offers a good combination of mechanical properties, corrosion resistance, and biocompatibility. In some industrial applications, it is necessary to proceed to finish treatments to extend the lifetime of the mechanical parts. In the present chapter, ball burnishing treatment is applied to improve the surface integrity of 316L since the performance behavior of parts is directly dependant on the surface properties of the used material. Both surface topography and surface microhardness of 316L after subjection to ball burnishing are studied. The number of burnishing passes is varied by up to five to investigate its effect on the results. Optical profilometer and atomic force microscopy (AFM) were used to analyze the surface roughness and surface topography texture while measurements of microhardness Vickers were proceeded to investigate the changes in surface hardening.

Keywords: 316L, ball burnishing, surface topography, microhardness

1. Introduction

Austenitic stainless steels, particularly 316L grade, have received much attention because of their good mechanical properties and high corrosion resistance [1]. This material contains a maximum carbon content of 0.03 by weight, which provides an extra level of corrosion resistance as well as the high rate of weldability. Several domains, notably marine and petrochemical industry, architecture, chemical production, and also biomedical sector, use this stainless steel due to its superior tensile strength, fracture toughness, and good formability [2]. 316L is non-magnetic and has excellent biocompatibility, which makes him a good candidate in the production of biomedical parts such as knee joints of total hip replacements [3]. In addition, 316L became very attractive to the industry owing to its low-cost and easy fabrication.

In almost engineering applications, an important interest is directed to the aspect of surface as it strongly influences the functional properties of mechanical parts such as their corrosion resistance, tribological behavior, and fatigue durability. Most failures of manufactured parts initiate from the outer layers which are exposed to the environmental conditions of service [4]. Mechanical, metallurgical, or chemical changes are the most common causes of the initiation of alterations in the surface [5]. In the case of wear, repeated contact actions between surfaces lead to the abrasion and/or delamination of the superficial layer which causes a loss in material quantity as well as in wear resistance. This loss is also produced in the case

IntechOpen

of corrosion where chemical changes in the surface are provoked after the contact between the surface and the environment under which the material operates. As a result, of these phenomena, the properties of surfaces become poor and unacceptable to fulfill the intended requirements of service. Some examples of components because of surface damages are: (a) environmental stress cracking of plastics by some chemical environments [6], (b) turbine vane and blade material surface deterioration caused by erosion [7], (c) surface corrosion [8], etc. The surface quality of materials therefore greatly needs attention to guarantee a good longevity of manufactured products.

The surface integrity notion, as it is understood is manufacturing processes, was defined by Field and Kahles [9] as *the inherent or enhanced condition of a surface produced in machining or other surface generation operation*. This term concerns many parameters:

- Topological characteristics (surface roughness, geometric aspects...);
- Mechanical characteristics (microhardness, residual stresses...);
- Metallurgical characteristics (phase transformation, grains size, ...);
- Chemical characteristics (changes in the chemical composition of the surface, ...).

Among the aforementioned parameters, surface roughness and microhardness are the major ones influencing the functional properties of parts. By far, the two parameters are remaining extensively studied to achieve better surface integrity. Surface roughness is a measurement of surface texture. A lower surface roughness indicates a smaller contact area with other materials, which is advantageous to improve corrosion resistance, frictional resistance, and fatigue life. Generally, the high quality of surface roughness is highlighted by the low values of amplitude parameters of surface topography. These parameters clarify the aspect of the topography which is related to the distance of a point on the surface from the mean plane, i.e., it gives information about the height or depth of a surface [10]. Hardness is the ability of a material to resist deformation. It is commonly preferred to produce surfaces with high values of microhardness as it prevents failures by wear and fatigue.

One way of improving the surface roughness and microhardness of parts is by applying surface treatments during the finishing step. Ball burnishing is a common mechanical surface treatment that has been widely applied on engineering parts for the finishing of their functional surfaces. This post-machining process is based on causing plastic deformation of the superficial layers through compressing a hard ball on the surface of the workpiece (**Figure 1**). As the ball is continuously moving, it transfers a material flow from peaks to valleys of superficial asperities. As a result, surface irregularities are reduced and compressive residual stresses are induced in the deformed layer. These two simultaneous actions improve the physical and mechanical characteristics of the surface which becomes smoother and also harder. Ball burnishing is easy, simple, and fast process which enhance the long-term properties of materials with low energy and almost no environmental pollution.

At present, there are rich literature sources about the effect of ball burnishing on surface roughness and microhardness of materials and also on the service performance of manufactured parts. The positive effect of this treatment in reducing the surface roughness [12–16] and raising the microhardness [12, 15–18] was widely reported. As a result, of these changes caused in surface characteristics, wear delamination was restricted as the interlocking movements of micro-irregularities were limited during friction [4, 19]. Fatigue resistance, yield and tensile strength, and also corrosion resistance were improved [20, 21].

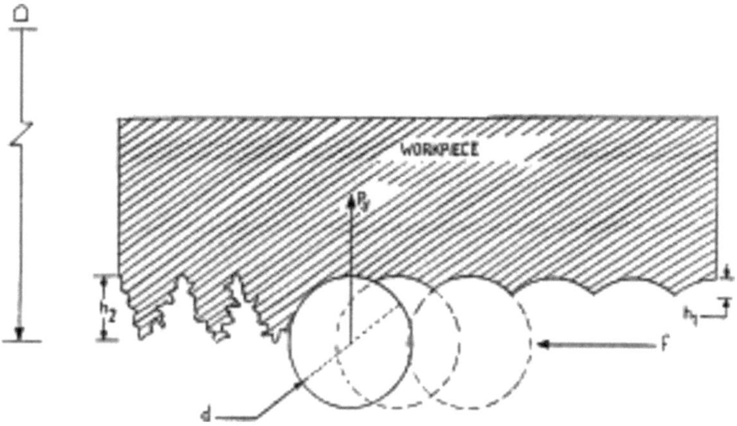

Figure 1.
Ball burnishing concept [11].

High surface finishing after ball burnishing is dependant on whether appropriate parameters of the process were well chosen or not. While the penetration depth and the initial state of the surface play a secondary role in obtaining good surface integrity, other parameters such as the burnishing force, the feed rate, and the number of passes contribute fundamentally to the final aspect of the treated surface [22]. Thus, it is very necessary to choose the right combination of process parameters and to master their effect on the surface integrity.

This research tackles the surface integrity of 316L after being subjected to the ball burnishing process. The effect of the number of burnishing passes, as an important process parameter, will be investigated. The results will be analyzed in terms of surface texture and microhardness after processing. At the sight of the results, an appropriate combination between burnishing force, feed rate, ball size, and a number of passes shall be proposed to execute the operation according to the right objective. This is important for 316L to confer its parts the special properties intended in the different industrial applications.

2. Experimental

2.1 Ball burnishing treatment

In this study, 316L stainless steel was used as workpiece material. The chemical composition of the alloy was determined as: the wt% is 0.02% C, the wt% is 16.64% Cr, the wt% is 10.35% Ni, the wt% is 2.03% Mo, and the rest is Fe. In the first step, a pre-machining operation was applied on a TOS TRENSIN machine to prepare the surface to be treated. Ball burnishing was then proceeded on the prepared surface with fixing the parameters regrouped in **Table 1**. The mentioned parameters were fixed after optimization of the process using the response surface "RSM" method based on BOX-Behnken design. The methodology of the optimization, as well as the

Parameter	Burnishing force (P_x)	Feed rate (f)	Ball diameter (D_b)
Value	87.1 N	0.18 mm/tr	13 mm

Table 1.
Burnishing parameters.

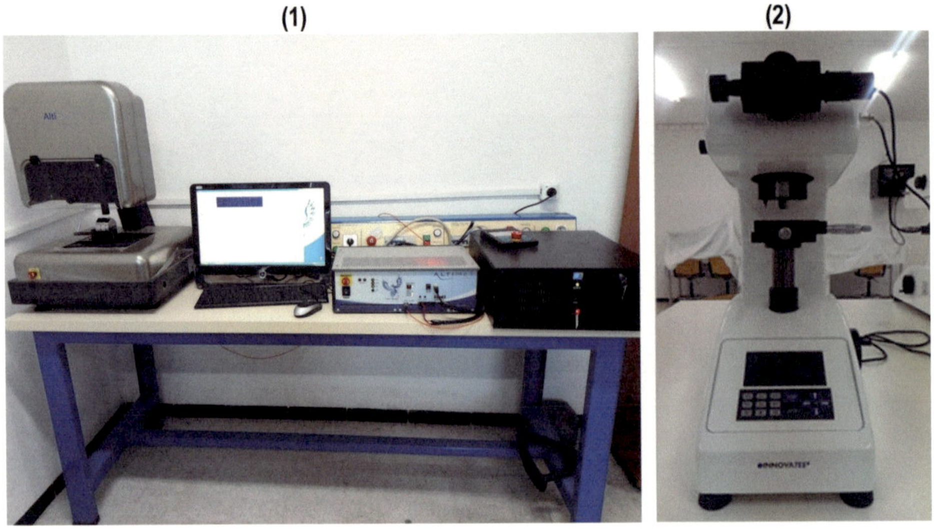

Figure 2.
Altisurf500 profilometer (1); Innovatest microdurometer (2).

analysis, were precisely described in our previous work [23]. After fixing the cited parameters, a number of passes (i) was varied by up to i = 5.

2.2 Characterization

The surface roughness of burnished surfaces was measured using Altisurf500 profilometer (**Figure 2**(1)). For each sample, an area of 3 × 2 mm^2 was scanned with a cutoff length of 0.8 mm. Altimap software was used to extract the following surface texture parameters. Atomic force microscopy (Bruker Dimension) was used to analyze the 3D topography of some selected surfaces.

Innovatest microdurometer, showed in **Figure 2**(2), was used to measure the surface microhardness of pre-machined and burnished samples. All tests were carried out using a load of 100 kgf applied for 10 s. The average diameters of five indentations were calculated to get reliable data.

3. Results and discussion

3.1 Effect of ball burnishing on the surface topography

The values surface topography parameters of the height of the five burnished samples, in comparison to the turned one, are regrouped in **Table 2**.

As a consequence of the ball burnishing process, the root means square height of the surface Sq was decreased in all samples as compared to the untreated one. During ball burnishing, the first two passes significantly reduced the Sq parameter from 2.212 μm to 0.715 and 0.729 μm respectively. After three passes (i = 3), the Sq still decreased achieving 0.583 μm. However, further augmentation in the number of passes provoked an increase in the Sq parameter which can attain a value higher than the initial one after five passes (i = 5). The turned surface is characterized by a more negative skewness Ssk and higher kurtosis Sku than all burnished samples. The highest skewness and the lowest kurtosis are registered in the surface burnished with three passes (i = 3/Ssk = −0.784 Sku = 2.77). The other amplitude parameters (Sp, Sv, Sz, and Sa) of all samples decreased after ball burnishing. These parameters follow

Surface topography parameters of height		Sample					
		Turned	**Burnished (i = 1)**	**Burnished (i = 2)**	**Burnished (i = 3)**	**Burnished (i = 4)**	**Burnished (i = 5)**
Root mean square height	Sq [μm]	2.212	0.715	0,729	0.583	0.978	2.62
Skewness	Ssk	−1.121	−0.758	−0,856	−0.784	−0.799	−1.09
Kurtosis	Sku	5.83	3.62	3.48	2.77	5.34	3.24
Maximum peak height	Sp [μm]	5.89	3.37	3.37	1.35	2.83	3.60
Maximum pit height	Sv [μm]	9.52	5.76	3.85	2.16	3.79	7.50
Maximum height	Sz [μm]	15.41	9.14	7.23	3.51	6.62	11.1
Arithmetic mean height	Sa [μm]	1.552	0.574	0.590	0.450	0.733	2.12

Table 2.
Comparison of height surface topography parameters of ground and ball-burnished samples.

the same tendency with respect to the number of passes as the Sq parameter. Hence, it can be noted that three passes (i = 3) are the most appropriate if surface topography is aimed to be improved. Further augmentation in this parameter can lead to the deterioration of surface quality, which is indicated by the increase in the amplitude parameters. The main objective of ball burnishing is the reduction of the heights of surface irregularities. Effectively, this objective was reached because the results show that the height parameters of the surface structure were reduced by more than three-fold, which is indicated in the results listed in **Table 2**.

The sample burnished with three passes (i = 3) shows the best height surface topography parameters. As a result, this surface is selected to be studied in terms of the other surface topography parameters. **Table 3** regroups the measured parameters of this surface as compared to the turned sample. Comparing these results, significant differences in the measurements can be highlighted between the turned and the burnished surface. After ball burnishing, the areal material ration Smr was significantly increased while the Smc and Sxp indicators were reduced. The Smr parameter of the turned sample was very low indicating a high peaky topography. After the application of ball burnishing, the value of Smr was sharply increased which impacts positively on the wear properties of the material. Indeed, a good bearing ratio indicates a good bearing capacity which improves the tribological behavior of the workpiece [10]. The Sxp parameter was reduced indicating reducing in surface roughness [24].

The spatial parameter Str of the burnished sample was almost similar to that of the turned sample indicating the micro-anisotropic texture of both surfaces. The micro-anisotropy is a natural result of the machining process [10]. In turning, and similarly burnishing, the single point cutting tool will generate a high degree of anisotropy to the machined surface. The std. parameter is used to indicate the marked direction of the surface texture for the y-axis, which means indicating the lay direction of the surface [25]. This parameter is applicable only for surfaces which does not have a uniform texture, i.e., when the Str > 0.5. It can be observed from **Table 3** that both turned and burnished surfaces have a lower value of Str < 0.5, which means that both surfaces have a pronounced lay pattern. The Std parameter gives the direction angle of the texture, which in the present results has increased from 72.01° for the turned sample to 106° after burnishing.

Surface topography parameters			Sample	
			Turned	Burnished (i = 3)
Functional	Areal material ratio	Smr [%]	0.826	28.8
	Inverse areal material ratio	Smc [μm]	2.263	0.660
	Extreme peak height	Sxp [μm]	6.045	1.47
Spatial	Texture-aspect ratio	Str	0.292	0.312
	Texture direction	Std [°]	72.015	106
Hybrid	Root mean square gradient	Sdq	0.070	0.159
	Developed interfacial area ratio	Sdr [%]	0.247	1.11
Functional (volume)	Material volume	Vm [mm^3/mm^2]	1.088×10^{-4}	1.95×10^{-5}
	Void volume	Vv [mm^3/mm^2]	0.00237	0.00068
	Peak material volume	Vmp [mm^3/mm^2]	1.088×10^{-4}	1.95×10^{-5}
	Core material volume	Vmc [mm^3/mm^2]	0.00168	0.0005
	Core void volume	Vvc [mm^3/mm^2]	0.00196	0.000588
	Pit void volume	Vvv [mm^3/mm^2]	4.096×10^{-4}	9.22×10^{-5}
Functional (stratified surfaces)	Core roughness depth	Sk [μm]	3.118	1.26
	Reduced summit height	Spk [μm]	1.849	0.302
	Reduced valley depth	Svk [μm]	3.613	0.722
	Upper bearing area	Smr1 [%]	14.34	6.81
	Lower bearing area	Smr2 [%]	77.61	86.5

Table 3.
Comparison of surface texture parameters of turned and ball-burnished surfaces.

All functional (volume) parameters were significantly reduced as a result of the ball burnishing process. The decrease in the material volume Vm indicates that an important part of surface irregularities was eliminated while the decrease in the void volume Vv refers to the elimination of valleys. This is also evident from the diminish of the other functional parameters Vmp and Vvc. The Vvv parameter characterizes the volume of fluid retention in the deepest valleys of the surface. Although this indicator was reduced for the burnished surface, this is not significant as this parameter is not affected by wear processes [26]. The wear resistance of components is directly related to the functional (volume) parameters and the enhancement resulting after BB impacts positively on reducing the quantity of material exposed to wear during the functioning of the workpiece.

For the functional parameters (stratified surfaces), all the parameters were reduced in the case of the burnished surface. The only exception is for the parameter Smr2. The lower value of Sk is desired for better sliding contact between contact surfaces while the decrease in the Spk parameter means that the volume of the material which is likely to be removed during the running in of the component was considerably restricted [26].

Based on the previously cited results, the effect of ball burnishing on surface topography can be remarked. As a consequence of the ball burnishing process, the functionality surface topography of 316L was efficiently improved which is characterized by the advantageous micro-geometric changes, namely: surface smoothness, elimination peaks and valleys and reduced peak heights and trough depths. The effect of a number of burnishing passes was also highlighted. It can be concluded that when the ball passes repeatedly over the surface of 316L, it deforms more asperities and produce smoother surface. However, this repetition should be limited 3 times to have the most improved surface, otherwise, surface flaking occurs due to excessive plastic deformation on the same surface layers [4].

Figure 2 represents isometric views of the selected burnished sample with three passes (i = 3) which showed the best-enhanced surface topography parameters. According to the 3D images of the untreated surface (**Figure 3**(1)), we can notice that it is characterized by higher peaks and deeper valleys compared to the burnished surface (**Figure 3**(2)). Hence it can be confirmed that the burnishing treatment by applying three passes produced a smoother surface. The visible scratches on the turned surface are due to the machining process which generates significant roughness (Ra = 134 nm μm and Rq = 172 nm). After burnishing with three passes over the surface of the 316L, the scratches as well as the peaks have almost disappeared, which reduces the roughness Ra to 14.1 nm and Rq to 18.3 nm, i.e., a decrease by 89.4% and 68.3% respectively.

3.2 Effect of ball burnishing on the surface microhardness

The effect of ball burnishing as well as the number of passes on surface microhardness Hv of the surfaces is presented in **Figure 4**. It can be remarked that all burnished surfaces show higher values of microhardness in comparison to the turned sample. This indicated the high efficiency of BB process in hardening the surface of

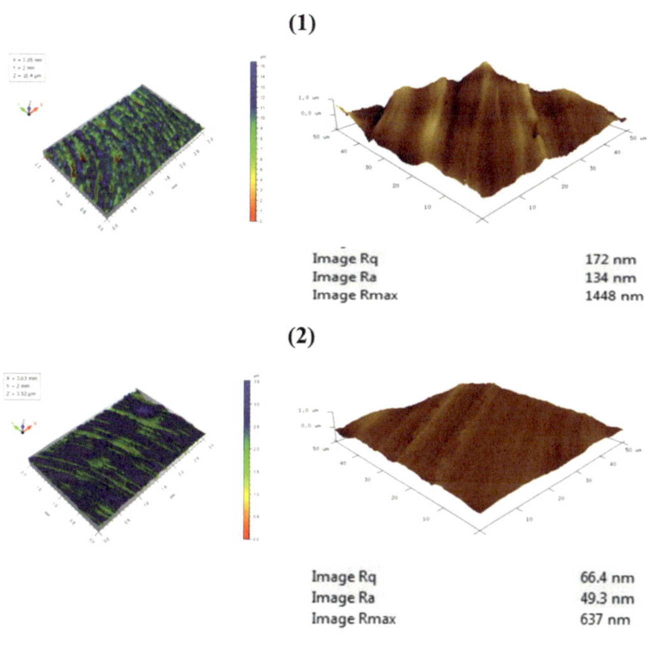

Figure 3.
3D images of (1) untreated surface and (2) burnished sample with i = 3.

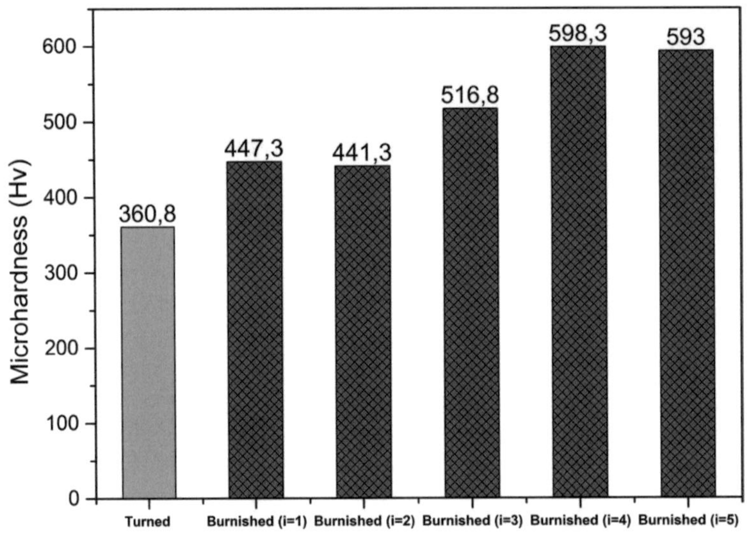

Figure 4.
Microhardness values of untreated and burnished surfaces.

316L. The impact of increasing the number of passes is manifested by a variation in the final microhardness. As a result, to one pass during ball burnishing (i = 1), an increment of approximately 90 Hv was obtained. While the increase in the number of passes to i = 2 did not really cause a remarkable change in the microhardness, the application of 3–5 passes generates a very significant hardening. The application of three passes has led to the increase in microhardness by 150 Hv while the most significant improvement in microhardness was recorded when four passes (i = 4) were applied. A similar hardening effect was caused after the application of five passes (i = 5) which is characterized by reaching a value of 593.0 Hv.

The impact of burnishing on the microhardness is interpreted by the plastic deformation which produces a structure with condensed grains and generates residual stresses loading the surface in compression. However, and contrary to the results of roughness, the increase in the number of passes does not cause any negative effect even if the ball passes 5 times successively over the same surface. Although, several works stipulate that repeating the burnishing operation several times, especially more than 3 times, destroy the surface because the surface is already saturated. We can explain our result by the high feed (0.18 mm/rev) and the low force (80 N) applied during the five passes. Indeed, since the ball is lowly loaded and moves quickly, it deforms more areas after each pass without affecting the already deformed areas. In other words, the repetitive passage of the ball over the same area will not have a detrimental effect since it does not have either the great force or the time sufficient to penetrate the surface and destroy the previously deformed layers.

4. Conclusions

In this study, the surface integrity of 316L stainless steel after ball burnishing was investigated. Based on the obtained results, the following conclusions can be drawn:

- The surface topography of 316L stainless steel was successfully improved after the ball burnishing process. The root means square Rq was reduced to less than 0.1 μm. Almost all the surface texture parameters were reduced after the application of ball burnishing treatment, which is an important advantage if the functional properties, such as the wear resistance, is aimed to be improved.

- The surface microhardness of the studied alloy was efficiently raised as a result of the ball burnishing process. An increment in microhardness by up to 150 Hv was recorded indicating the work hardening effect induced during the process.

- The number of passes is found to be an important parameter that influences on both surface topography and microhardness. The previously cited results are obtained in condition to the application of three number of passes. Further increase in the number of passes can lead to the deterioration of the surface.

Acknowledgements

The authors thank the response of Mechanics and Structures (LMS) Research Laboratory, University of 8th May 1945, Guelma, Algeria, for their involvement in our study.

Conflict of interest

The authors declare no conflict of interest.

Author details

Selma Attabi[1*], Abdelaziz Himour[1*], Lakhdar Laouar[2] and Amir Motallebzadeh[3]

1 Metallurgy and Materials Engineering Laboratory (LMGM), University Badji Mokhtar of Annaba, Annaba, Algeria

2 Laboratory of Industrial Mechanics (LMI), Badji Mokhtar—University of Annaba, Annaba, Algeria

3 Koç University Surface Science and Technology Center (KUYTAM), Istanbul, Turkey

*Address all correspondence to: attabi.selma@gmail.com and ahimour@yahoo.fr

References

[1] Fellah M, Labaïz M, Assala O, Iost A, Dekhil L. Tribological behavior of AISI 316L stainless steel for biomedical applications. Tribology—Materials, Surfaces and & Interfaces. 2013;7:135-149. DOI: 10.1179/1751584X13Y.0000000032

[2] Saboori A, Alberta A, Giulio M, Sara B, Mariangela L, Paolo F. Microstructure and mechanical properties of AISI 316L produced by directed energy deposition-based additive manufacturing: A review. Applied Sciences. 2020;**10**(9):3310. DOI: 10.3390/app10093310

[3] Chintan M. A review on parameters affecting properties of biomaterial SS 316L. Australian Journal of Mechanical Engineering. 2020:1-11. DOI: 10.1080/14484846.2020.1752975

[4] Attabi S, Himour A, Laouar L, Motallebzadeh A. Effect of Ball Burnishing on Surface Roughness and Wear of AISI 316L SS. Journal of Bio- and Tribo-Corrosion. 2021;7:7. DOI: 10.1007/s40735-020-00437-9

[5] Davim JP. Surface integrity in machining. In: Surface Integrity—Definition and Importance in Functional Performance. London: Springer; 2010. pp. 1-35. DOI: 10.1007/978-1-84882-874-2_1

[6] Characterization and Failure Analysis of Plastics. Materials Park, OH: ASM International; 2003. Available from: http://asmcommunity.asminternational.org/portal/site/www/AsmStore/ProductDetails/?vgnextoid=66b975dfae0f8110VgnVCM100000701e010aRCRD

[7] Sotnikov AA, Stepanov VN, Livshits AM, Bukchin SM. Improvement of turbine blade systems to reduce cavitation erosion. Power Technology and Engineering. 1994;**28**(12):746-750

[8] Masakazu O, Hiroshi M, Kazuyuki M, Tomoyoshi O, Masayuki U. Surface corrosion of various kinds of metals by atmospheric pollution. Bulletin of Hiroshima Kokusai Gakuin University. 2005;**38**:69-83

[9] Field M, Kahles JF. The surface integrity of machined and ground high strength steels. DMIC Report. 1964;**210**:54-77

[10] Waikar RA, Guo YB. A comprehensive characterization of 3D surface topography induced by hard turning versus grinding. Journal of Materials Processing Technology. 2008;**197**(1-3):189-199. DOI: 10.1016/j.jmatprotec.2007.05.054

[11] Hassan AM, Al-Bsharat AS. Influence of burnishing process on surface roughness, hardness and microstructure of some non-ferrous metals. Wear. 1996;**199**:1-8

[12] Revankar GD, Shetty R, Rao SS, Gaitonde VN. Analysis of surface roughness and hardness in ball burnishing of titanium alloy. Measurement. 2014;**58**:256-268

[13] Dzierwa A, Gałda L, Tupaj M, Dudek K. Investigation of wear resistance of selected materials after slide burnishing process. Eksploatacja i Niezawodnosc–Maintenance and Reliability. 2020;**22**(3):432-439

[14] Świrad S, Wydrzynski D, Nieslony P, Krolczyk GM. Influence of hydrostatic burnishing strategy on the surface topography of martensitic steel. Measurement. 2019;**138**:590-601

[15] Teimouri R, Amini S, Bami AB. Evaluation of optimized surface properties and residual stress in ultrasonic assisted ball burnishing of AA6061-T6. Measurement. 2018;**116**:129-139

[16] Jerez-Mesa R, Travieso-Rodríguez JA, Landon Y, Dessein G,

Lluma-Fuentes J, Wagner V. Comprehensive analysis of surface integrity modification of ball-end milled Ti-6Al-4V surfaces through vibration-assisted ball burnishing. Journal of Materials Processing Technology. 2019;**267**:230-240

[17] Dimitrov DM, Slavov SD, Dimitrov Z. Experimental research on the effect of the ball burnishing process, using new kinematical scheme on hardness and phase composition of surface layer of AISI 304L stainless steel. MATEC Web of Conferences. 2017;**112**:02001

[18] Basak HM, Ozkan T, Toktaset I. Experimental research and ANN modeling on the impact of the ball burnishing process on the mechanical properties of 5083 Al-Mg material sheets through vibration-assisted ball burnishing. Kovove Materialy. 2019; **57**(1):61-74

[19] Revankar GD, Shetty R, Rao SS, Gaitonde VN. Wear resistance enhancement of titanium alloy (Ti–6Al–4V) by ball burnishing process. Journal of Materials Research and Technology. 2017;**6**:1, 13-32

[20] Avilés R, Albizuri J, Rodríguez A, De Lacalle LL. Influence of low-plasticity ball burnishing on the high-cycle fatigue strength of medium carbon AISI 1045 steel. International Journal of Fatigue. 2013;**55**:230-244

[21] Sadeler R, Corak S, Atasoy S, Bulbul F. Evaluation of an electroless nickel-boron (Ni-B) coating on corrosion fatigue performance of ball burnished AISI 1045 steel. Kovove Materialy-Metallic Materials. 2013;**51**(6):351-356

[22] Laouar L, Hamadache H, Saad S, Bouchelaghem A, Mekhilef S. Mechanical surface treatment of steel-Optimization parameters of regime. Physics Procedia. 2009;**2**:1213-1221

[23] Attabi S, Himour A, Laouar L, Motallebzadeh A. Mechanical and wear behaviors of 316L stainless steel after ball burnishing treatment. Journal of Materials and Research Technology. 2021;**15**:3255-3267

[24] Marteau J, Bigerelle M, Mazeran P-E, Bouvier S. A Relation between roughness and processing conditions of AISI 316L stainless steel treated by ultrasonic shot peening. Tribology International. 2015;**82**:319-332

[25] Dong WP, Sullivan PJ, Stout KJ. Comprehensive study of parameters for characterizing three dimensional surface topography. IV. Parameters for characterizing spatial and hybrid properties. Wear. 1994;**178**:45-60

[26] Blateyron F. The areal field parameters. In: Leach R, editor. Characterization of Areal Surface Texture. London: Springer; 2013. DOI: 10.1007/978-3-642-36458-7_2

Chapter 7

Welding Based Additive Manufacturing: Fundamentals

Maruthasalam Sowrirajan, Selvaraj Vijayan and Munusamy Arulraj

Abstract

Additive Manufacturing (AM) has drawn abundant attention over the past decades in the manufacturing and fabrication industries, especially to make part models and prototypes. This chapter introduces a potential welding based AM process called Wire Arc Additive Manufacturing (WAAM) for the fabrication of near-net shaped metal components including stainless steel components. To start with traditional AM processes, various fundamental traditional AM for the fabrication of components have been presented. Wire Arc Additive Manufacturing (WAAM) has been explained with its variants, synonyms, different welding processes to suit WAAM particularly to weld stainless steel metal; primary process selections for working with WAAM, important metals, and alloys that could be used in WAAM have been elaborated. A case study for WAAM fabrication of AISI 316 L stainless steel plate is included to introduce the fabrication of metal components using WAAM. Further, the most common defects which possibly play a vital role in WAAM components fabrication and a few of the future challenges regarding WAAM development are discussed. Fundamental information covered in this chapter could be more beneficial to beginners for the understanding of WAAM process generally including stainless steel component fabrication in a lucid tactic.

Keywords: additive manufacturing, welding, Wire Arc Additive Manufacturing, fabrication of stainless steel components, mechanical and metallurgical properties, defects, quality aspects

1. Introduction

In recent days, Additive manufacturing (AM) is often called as 3D printing and this is basically a transformative method to the conventional industrial production that takes us to the creation of light as well as strong components simply by adding the material layers [1]. This is generally called as additive layer manufacturing (ALM) because the final component is fabricated by depositing materials, usually layer over a layer. This ALM is shown in **Figure 1**. Any shaped 3-dimensional objects may be fabricated with the help of this AM concept. This is a dominant process nowadays occupied by the majority of industries over the subtractive manufacturing methods such as machining and cast forming. This is because the fabrication of components using this AM looks effortless compared with subtractive processes [2]. AM is believed to be the future of many industries, simply not

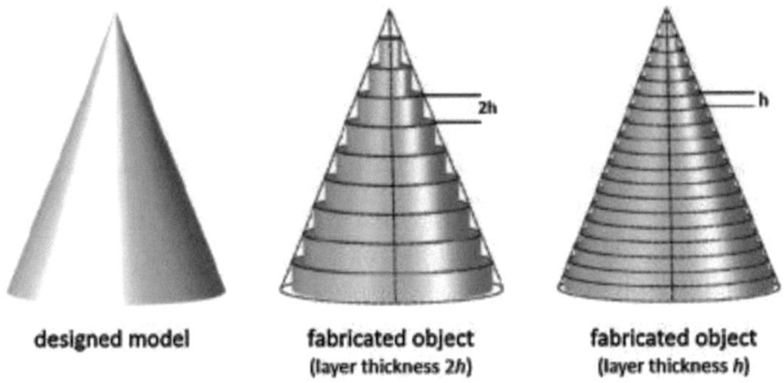

Figure 1.
Concept of additive layer manufacturing (ALM).

only in manufacturing factories but also in the construction fields. This 3D Printing technology is now widely used in the field of constructions.

1.1 Additive manufacturing processes

Selection of an appropriate process is very important among the AM processes for successful fabrication. Generally, there are number of different methods available in standard to meet the AM processes and they are listed below.

1. Binder jetting

2. Directed energy deposition (DED)

3. Material extrusion

4. Powder bed fusion

5. Sheet lamination

6. Vat polymerization

7. Directed energy deposition

To start with Binder jetting, it actually uses a 3D printer head and moves in three standard coordinates (x, y, and z axes) to print all the layers of powder material and also a liquid form one known as the binder for adhesive purposes. DED shall be employed with almost entire materials including metals, ceramics, and polymers irrespective of the weight of material for the fabrication of final components. Electric arc, laser, and electron beam shall be used for the melting of solid wire material or feedstock filaments or metal powder to deposit the melted material as layers. Wire Arc Additive Manufacturing (WAAM) is one of an arc welding based AM. Material extrusion involves the spooled polymers which may be extruded or drawn into a heated nozzle. This also builds melted material in the form of layer over a layer and finally, these layers stick either because of temperature control or chemical bonding. Powder bed fusion technology generally involves the formation of melts or partially melts layers of material cleaned by blasting away the excess powder. Many AM techniques follow this concept includes Direct Metal Laser Melting (DMLM), Direct Metal Laser Sintering (DMLS), Electron Beam Melting (EBM), Selective Laser Sintering (SLS), and Selective Heat Sintering (SHS).

Thin sheets of material are laminated over one another together to fabricate a single 3D component is known as simply the sheet lamination. Laminated Object Manufacturing (LOM) and Ultrasonic Additive Manufacturing (UAM) are available technologies for sheet lamination. Vat polymerization customs a liquid resin photopolymer vat to produce a piece [3]. Layers of materials are cured by ultraviolet light by mirrors is called photopolymerization.

Even though the above-listed processes are available for AM, these processes would fall under any one type of technology from three technologies. The first technology is called *sintering* in which the powder form of material used for the fabrication is heated up to the stage well before liquefied for the creation of complex high-resolution components. The second one involves in complete *melting* in which the powder material used for the fabrication is fully melted for the creation of the part. Generally, the laser is used for the action of this melting here. The another third type of technology is referred to as *stereolithography*, in which a process of photopolymerization is employed, therefore, an ultraviolet laser is fired and directed to photopolymer resin to make torque-resistant parts that endure temperatures at both extremes [4]. These processes and technologies are used for the successful 3D printing of components of various materials.

In general, the powder-based additive manufacturing process is so familiar and commercially readily available in most of the nearby industries that fabricate the components that are smaller in size. Our matter of interest WAAM is the wire-based welding process that has emerged to date as a potential candidate for comparatively large AM metal components. The chapter provides the description on WAAM technology and its possible classification while focusing on stainless steel and various synonyms of WAAM being used worldwide have been presented. Various arc-based welding processes have been discussed and a case of stainless steel plate fabrication is given in detail. Various weld defects in WAAM fabrication are presented and a few of the future works while focusing on the fabrication of stainless steel components are presented.

1.2 Wire Arc Additive Manufacturing (WAAM) technology

The WAAM is a typical AM technology under DED technologies. Generally, in DED, the energy sources such as laser, electric arc, and electron beam are used to melt the feed material in the form of either powder or solid metal wire. Making use of the welding process would enable the heat energy of the process to focus on the metal to be melted directly. This kind of welding based melting and depositing the metal layers additively in order to achieve the final shape and size of the component is known as Wire Arc Additive Manufacturing (WAAM). Therefore, this WAAM particularly falls as one of a DED technique. This WAAM uses an appropriate welding process for melting the metal wire electrode and depositing the melted metal layers over another as in the case of Additive Manufacturing (AM) for the fabrication of metallic components and this is now being one of the fastest and growing research and development field among the almost entire area of engineering fabrication. WAAM generally uses any of the arc welding process power sources along with manipulators for automation to deposit 3D shaped metal components. Manipulations are familiarly being done by CAD/CAM based softwares. Robotic welding equipment is well suited for WAAM because it is easy to get required shape automation/manipulations. The shape of the component being manufactured is achieved by planning the path of the depositions well. This process most normally uses solid wire as a feed material. This solid wire gets melted by an electric arc and moves according to the well-preplanned paths to deposit the metal component in the desired shape. Therefore, the melted metal is deposited layer over a layer to create the required component [5]. **Figure 2** shows a welding based WAAM process for the fabrication of stainless steel plate components in which the arc generated in any

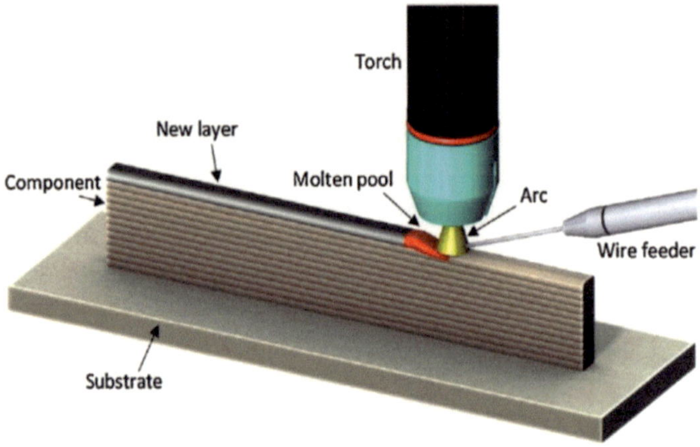

Figure 2.
Illustration of WAAM.

of the arc welding process is directed to the metal wire coming out from the feeder and the arc melts the metal wire thus deposition will be made over a substrate. Thus, the number of required deposition layers would be allowed up to attaining the required shape and size of the component [6].

WAAM illustration may look very simple but in nature, it is not a simple process to carry out. For a WAAM process, the process parameters of the welding process are considered to control to yield desired deposition weld bead geometry there by the final shape of the component is achieved as a single component by arranging weld bead/layer over weld beads/layers. Much care should be given to the welding parameters selected more than the welding operation as these parameters are capable of affecting the quality and shape of the component being fabricated [7]. Therefore, WAAM needs a skilled operator for successful fabrication.

1.2.1 Classification of WAAM

The WAAM is a hot developing topic yet and is not classified so far clearly. But in this text, an attempt has been given for the classification of WAAM processes based on the familiar arc welding processes alone employed for stainless steels and proved to be fit for WAAM. Therefore, WAAM is here classified into,

1. Gas Metal Arc Welding (GMAW) based WAAM

2. Gas Tungsten Arc Welding (GTAW) based WAAM

3. Plasma Arc Welding (PAW) based WAAM

Selection of the welding process is based on the operator's requirement. For example, if the priority is given to a high deposition rate, GMAW process may be selected compared to GTAW even if the quality and process stability are comparatively less. If the operator is looking for an electric arc with very high energy density then PAW may be preferred that actually enables increased travel speeds and good quality weld deposits with minimum weld distortion [8].

1.2.2 Synonyms of Wire Arc Additive Manufacturing

The WAAM is presently well-known synonym for Directed Energy Deposition-Arc (DED-arc) that enables the need well to fabricate, any of the

S. no	Terms used	Used organizations
1	Welding based deposition	Southern Methodist University
2	3D Welding	Korea Institute of Science and Technology, University of Nottingham
3	Welding based rapid prototyping	University of Kentucky
4	Near-net shape manufacturing	Tufts University
5	Shape deposition manufacturing	Carnegie Mellon and Stanford University
6	WAAM	Cranfield University, University of Wollongong
7	Welding based AM	National Institute of Technology Tiruchirappalli, Coimbatore Institute of Engineering and Technology

Table 1.
Different terms used for WAAM by others worldwide.

engineering structures with very near net shapes but not to the exact shapes. As known, Wire Arc Additive Manufacturing (WAAM) is a welding based fabrication process used to 3D print metal parts. The WAAM is also used to repair the metal parts as well [9]. Cold Metal Transfer mode in the welding process is now extended particularly to avoid the problems due to heat accumulation during weld layers deposition is simply now announced as CMT-WAAM by Fronius [7], a company working vigorously in worldwide welding. Sometimes because of the reason of using solid metal, it may also be referred as Metal Additive Manufacturing (MAM) for both AM as well as welding based AM process. Hence, the WAAM synonyms include the terms such as DED-wire process, 3D Printing process, Wire and Laser AM (WLAM), Electron Beam Freedom Fabrication (EBFF), Metal AM process, Welding based AM process, and CMT-WAAM process. Many have referred to this welding based AM process by various terms [9]. They are clearly represented in **Table 1**.

2. Welding processes for additive manufacturing

The WAAM generally requires a welding process giving high deposition rates without quality and size limitations so that WAAM could become the best alternative for conventional AM processes, especially when dealing with medium and large size component fabrication. The increasing interest over the world in WAAM is mainly because of the advances in robotic controls and modern electronic control circuit systems for different welding machines [10]. The modern methods of controlled short arc processes with reduced heat input like a process called Cold Metal Transfer (CMT) technology were particularly developed for GMAW process and now has been developed by Fronius International, a company works worldwide especially for the WAAM in the name of CMT based GMAW. Tungsten Inert Gas (TIG) welding process and Plasma Arc Welding (PAW) process could be used due to the separation of mass and energy input. Also, apart from the standard welding processes, laser-assisted welding has also been available for this type of AM. The following welding processes have been experimented especially for fabrication worldwide with the focus on the best fabrication practice for fabricating the stainless steel components by WAAM [11].

1. Plasma arc welding (PAW)

2. Gas metal arc welding (GMAW)

3. Gas tungsten arc welding (GTAW)

4. Submerged arc welding (SAW)

5. Skeleton arc welding

6. Laser-assisted welding

2.1 Plasma arc welding (PAW)

Plasma arc welding process shall be employed for AM using filler material for the fabrication in either powder form or wire form. This may familiarly be called as Plasma Transferred Arc Welding (PTAW), an important variant of PAW. This welding process is found to be the best process for the depositions to suit surfacing or cladding processes with several materials. The same deposition approach may also simply be extended to a very simple approach for AM as like multi-layers surfacing or cladding. This process is capable of separating the energy input and the weld metal flow entirely from the filler metal. This process is applicable to different type of materials.

2.2 Gas metal arc welding (GMAW)

GMAW is a more frequently employed famous welding process for many decades in metal surfacing, cladding, and additive manufacturing too. Thus, this process is very familiar nowadays for WAAM because of its suitability. In recent years, much focus has been given to the development of control systems for welding arc power and this leads to the development of new process variants such as CMT-GMAW [7]. Furthermore, these controlled short arc welding processes can be ranked high by innovative wire feed control to achieve control on droplets. Also, a variant called the multi-wire process has been developed particularly to improve the mass of the deposited metal.

2.3 Gas tungsten arc welding (GTAW)

This GTAW is familiarly known as Tungsten Insert Gas (TIG) welding. The important major benefit of this TIG welding process is that the separation between the filler metal and heat energy input is highly possible. The possibility for using more than a single wire during the process made TIG an well-suited process for WAAM. A variant of TIG called the pulsed TIG welding process has shown a very good melt pool shape thus to suit for additive manufacturing. Various properties of the AC-TIG and pulsed TIG welding processes are also under rigorous investigation to meet the WAAM process. It is believed that GTAW would be a great choice for WAAM in near future.

2.4 Submerged arc welding (SAW)

SAW can be a successful welding process for AM but positioning the flux plays an important role in this process because of this reason the flexibility of this process

Figure 3.
Skeleton arc WAAM.

is limited AM. For carrying the electrode and current, a welding gun and cable set are used and thus deliver the flux to the arc. The flux to be supplied depends upon the position of the gun. On the other hand, the torch is attached to the motor that is feeding wire and takes necessary actions regarding current to take tips for the reason of transmitting the welding current to filler wire. SAW would become complex by including additional devices like weavers, seam followers, and work rovers, etc. Because of these complex reasons SAW has not been identified with ease for the WAAM process.

2.5 Skeleton arc welding

Skeleton arc welding is said to be a distinct variant of WAAM. The final part would be fabricated by single dot depositions. The creation of skeleton-like structures in almost every component is possible in this method. These skeleton structures could be employed to make lightweight path-oriented designs into AM and are clever to make bionic-like structures easily. All the materials could be used in this type of WAAM provided the path planning is executed well. A typically employed WAAM by skeleton arc welding with the best path planning is represented in **Figure 3**.

2.6 Laser-assisted welding

Laser-assisted welding completely differs from the process of laser hybrid welding and this in fact typically practices the laser power lesser than 1 kilowatts and thus one of the possible benefits geometrical control of the weld deposit any few more benefits also possible. Moreover, the surface quality is also found to be enhanced while using less power during the operation. This laser-assisted welding will become the appropriate process soon preferably for WAAM.

3. Primary process selections for WAAM

Primary WAAM process selections are very much essential for confirming the rightness of the fabricated structures and this indeed requires characterization of the deposited material. Many of the mechanical, thermal, and metallurgical properties may differ in the end components because it is fabricated by using any of the welding processes. This fundamental classification may include the bases such as substrate material, electrode wire, welding processes, welding process variables, shielding gases, and automation systems for the motion [5]. Therefore, primary WAAM process selection includes the following essentials,

1. Parent plate material

2. Electrode wire

3. Welding processes

4. Welding process variables

5. Shielding gases

6. Automation systems

3.1 Parent plate material

The section of substrate material is so important in WAAM because the wettability of deposited weld beads to build the component solely depends upon the chemical composition of the substrate material over which the weld beads are deposited. If the case is of dissimilar metal, then laying the weld bead onto the dissimilar substrate is very difficult one to consider. Moreover, weld dilution of the first layer of the deposit must also be analyzed for this case. The stiffness and thickness of the substrate material will offer enough resistance to weld distortion. Therefore, a detailed study of substrate material used during the fabrication of metal components is much important.

3.2 Electrode wire

The electrode wire to be used for the component fabrication by WAAM is driven by technological aspects based on the welding process concerned. The selections of electrode wire during WAAM are considered instrumental to the fabricated components performance. The wire gauge and the number of wires to be fed to the electric arc at a few of the welding process parameters definitely affect the performance measures such as the deposition rate, and the heat transfer behavior in the weld pool. This can brought welding defects like lack of fusion. The defects in the wire surface, like diameter variations, cracks, and some scratches could result in porosity within the deposited component. It has been proved that making use of high-quality wire results the lesser porosity in the welds. To handle carefully these defects, wire selection is very much important in WAAM fabrication.

3.3 Welding processes

The welding processes considered with foremost importance for WAAM depending upon the feasibility and cost benefits may be among Gas Metal Arc Welding (GMAW) process, Gas Tungsten Arc Welding (GTAW) process, and Plasma Arc Welding (PAW) process. Few more welding processes have been also available as discussed in Section 2, but the suitability for WAAM is very high for the above said frequently used welding processes. The selection of a suitable welding process depending on the component to be fabricated is already discussed in Sections 1.2.1 and 1.2.2 above.

3.4 Welding process variables

Several welding process variables are available for different welding processes. These welding process variables are deciding the final quality of components fabricated by WAAM. Therefore, a key care to be taken for the selection of appropriate welding process as well as welding parameters. Sometimes these variables may show the combined effect when selected together. Familiar process parameters of welding processes to consider include welding current, voltage, open-circuit voltage, wire feed rate, welding travel speed, welding electrode, diameter, electrode angle, substrate material, electrode dip to substrate distance, and more. The operator has to take much care while selecting the combination of welding process variables to yield the best quality in the fabricated components [12]. Also, the care must also be given over the range over which each selected parameters are to be operated to achieve defect-free components.

3.5 Shielding gas

The purpose of using the shielding gas in welding based WAAM is to protect the molten metal during welding against surrounding atmospheric elements. An acceptable flow rate of shield gas is essential to cover the hot weld deposit in WAAM to separate the atmospheric gases and to prevent the action of detrimental oxide, nitride, and porosity formation and porosity. This shield zone could be taken to the side walls also during the fabrication of WAAM components. A very high flow rate could lead to reduced weld penetration. Most commonly used shielding gases are argon, carbon dioxide, helium, nitrogen, hydrogen, and mixers of these gases.

3.6 Automation systems

Automation is the heart to deliver WAAM fabrication from welding. Cartesian coordinates (x, y & z axes), robotic arm, and manipulator like kinematic machines have been available for providing necessary motions to the weld-torch for attaining build shaped structure. The repeatability with accuracy in such motions will lead to achieve desired geometrical, physical properties of the weldment. The high volume of material may be needed to be post-processed, if the motion system is not accurate. This would detrimentally decide the cost of WAAM. CNC machine movements have been employed for WAAM and are believed to be cost-effective approach and would be a hybrid process as the motion for building components and machining to achieve the final component shall be done on the same machine at a stretch.

4. Metals and alloys for Wire Arc Additive Manufacturing

WAAM processes exactly use available electrode wires of materials that are produced by the welding industries and readily available in the spooled form in alloys as feedstock to welding processes. The materials to choose for WAAM process are completely based on the application requirements. For example, some materials may be preferred by automotive industries, aerospace industries, and so on [5, 13]. To cover these application requirements, the following familiar metals are available for WAAM.

1. Nickel-based alloys
2. Aluminum alloys
3. Chromium alloys
4. Titanium alloys
5. Stainless steel alloys
6. Other pure metals

WAAM community suggests that the Nickel-based alloys as the most popular materials for the fabrication by welding process due to their weld ability and high strength at elevated temperatures. Nickel-based alloys are widely used in aerospace, petrochemical, chemical, and marine industries due to their exceptional strength and oxidation resistance at high-temperature ranges. In WAAM, alloys such as Inconel 718 and Inconel 625 have been proved to be feasible alloys so far for WAAM processing. Fabrication trials for many different series of aluminum alloys have been carried out already using Al-Cu (2xxx), Al-Si (4xxx), Al-Mg (5xxx) and also using many alloys. The WAAM is mainly a justifiable process for structures as the cost of manufacturing the small and simple aluminum alloy component by conventional machining processes is comparatively less. Though many of the Al alloys such as Al 7xxx and 6xxx are still challenging to weld for WAAM fabrication due to the reason of turbulent melt pool and defects, the rigorous activities are being undertaken regarding aluminum alloys because of the superior benefit that additively manufactured aluminum alloy parts have inferior mechanical properties compared to those machined from billet material. In order to achieve higher tensile strength, most of the as-deposited aluminum parts undergo post-process heat treatment to refine the microstructure. Compared to other alloys these alloys are cheaper and easily available, the WAAM of these components gaining enormous interest irrespective of fabrication complexity.

Additions of chromium to iron and aluminum as an alloying element have been to improve the anti-corrosion properties of the structure and could be a potential candidate for increasing the lifetimes of the various corrosive environment applications. Titanium alloys contain a mixture of pure titanium and other chemical elements and these alloys have high tensile strength and toughness at the wide range of temperatures. They are generally lighter in weight and pose superior anti-corrosion properties. However, the high cost and processing limitations make these alloys to use in superior areas such as military and aerospace applications, sports, and bio-medical industries. WAAM is highly attempted for the fabrication for these areas with cost benefits. Stainless steel originally referred to as rustless steel, is a group of ferrous alloys that contain a minimum of approximately 11% chromium, this has made stainless steel as rust-free together with heat-resistant properties. Different types of stainless steel generally include the elements such as carbon, aluminum, silicon, nitrogen, sulfur, nickel, titanium, copper, selenium, niobium, and molybdenum. Sensitization is a major problem in austenitic type of stainless steels comparing ferritic steels and martensitic steels. Apart from this, residual stress, metallurgical properties, weld distortion and thermal properties of the WAAM manufactured components are to be taken with much care for the successful lifetime of components. Other metals such as magnesium, bronze, and many more intermetallic structures could be possible with this WAAM process but care must be given to improve the properties of such fabricated structure components.

5. Fabrication of components using WAAM

WAAM has already proven that it could fabricate medium to large-sized components as well because of high deposition rate welding processes. Similar to AM technologies, WAAM could produce uniform, defect-free metal components are a key reason for the selection of WAAM. Moreover, WAAM is not simply a prototyping technology and most of the present day's attention is on using this WAAM as a transformation to a viable and cost-effective fabrication. Fabrication of a metallic plate component by Welding based Additive Manufacturing (WAM) is explained for an easy understanding of WAAM fabrication.

As an example, the low-carbon high-strength steel (AWS A5.28 ER110S-G) metal plate as shown in **Figure 4** is fabricated by WAM. The experimental apparatus is a GMAW based WAAM setup available at the Coimbatore Institute of Engineering and Technology, Coimbatore. The experimental process consists of a welding torch mounted on a linear manipulator system. An Ador makes GMAW power source based experimental setup was employed to deposit the metal over the parent plate. The feedstock material was AWS A5.28 ER110S-G filler wire, a low-carbon high-strength steel wire electrode with a diameter of 1.2 mm. The metal plate was built on a mild steel substrate plate. The GMAW parameters such as voltage, current, wire feed rate, gas flow rate, and welding speed were considered as the primary parameters and those values were kept as 20 V, 100 A, 4 m/min, 12 l/min, and 4 mm/s respectively for the fabrication of decided plate component. Totally 15 numbers of weld beads were deposited and this structure is capable of giving 10 × 100 mm desired dimensions plate component after a small machining operation. This is the reason why the WAAM component is famously described as a near-net component.

After the fabrication of this plate-shaped component, a lot of properties such as surface waviness, weld bead dimensions, height of each layers, and weld bead angle are evaluated essentially to speak about the geometrical specifications of fabricated component and similarly, mechanical properties such as tensile strength, ductility, fatigue strength, corrosion rate, hardness, and microstructural evaluations are essential to be carried out in order to meet the quality and quantity benefits of fabrications as well.

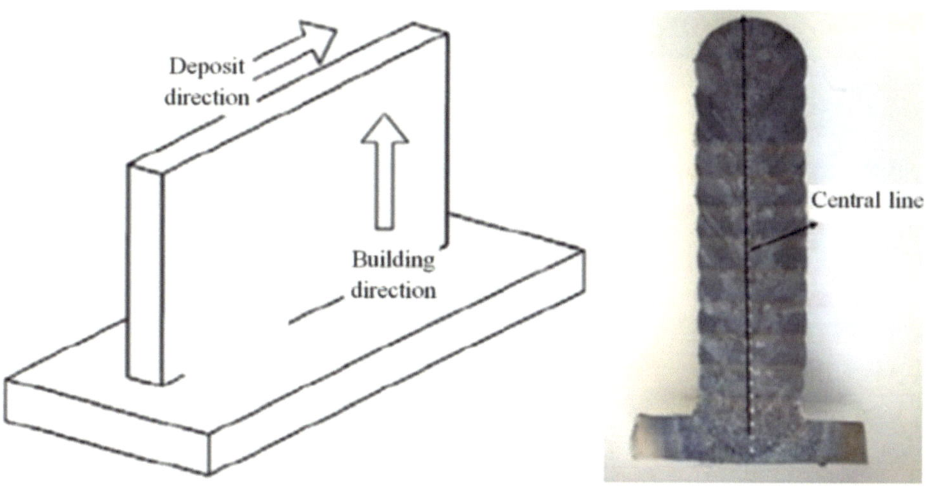

Figure 4.
Fabrication of components using WAAM.

6. Most common defects in WAAM components

It has been seen that WAAM is a welding based fabrication method and components are being made to any shape and size according to the requirements. Therefore, everyone might think that the properties of WAAM fabricated components would match with the conventional welding process. But this fact is not true because the welding process will be carried out for many purposes such as different weld joining and weld cladding, etc. Almost in all the like cases, it is obvious that the weld bead would be limited to only a single in most of the cases and if needed, in peculiar cases few weld beads would be needed to accomplish the process needs. The WAAM case especially while fabricating stainless steel components is entirely different because of the reason that the entire component is manufactured by depositing the weld beads. In this case, many more weld beads to be deposited especially in layers according to the component being fabricated [11]. Therefore, the weld beads may get different properties than the normal welding. There is a strong possibility of more defects in WAAM fabricated components.

Though the meaning of these weld defects in WAAM components are similar to the welding process, it should be understood that there is highly possible to receive more amount of weld defects here since many more required number of weld beads are accumulated together to form the final component. As the components are fabricated with defects by WAAM, it is very tedious to meet the quality and performance-wise factors with the components manufactured in conventional methods or traditional AM processes. Therefore, it is strongly recommended to look for options to reduce these defects during the fabrication of metal components in WAAM [14]. The most common possible few of the weld defects in WAAM based

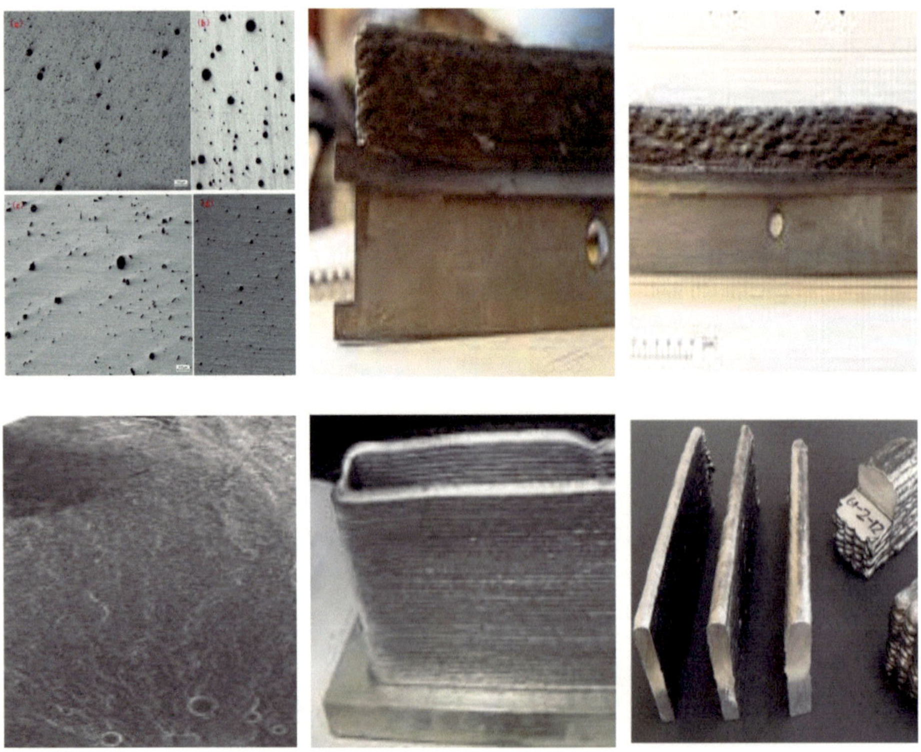

Figure 5.
Various defects in WAAM components.

fabricated components are listed below. Also, the weld defects that arise during WAAM fabrication are represented in **Figure 5**.

1. Porosity

2. High residual stress

3. Solidification cracking

4. Fatigue

5. Oxidation

6. Poor surface finish

7. Delamination

8. Deformation

7. Future challenges

Ready acceptance worldwide in many of the fabrication industries among the market for the welding based additive manufacturing of the numerous near net-shaped component fabrication is a noteworthy message to the research community for attending the forthcoming challenges associated with real-time fabrication of stainless steel components in WAAM. Some of the future challenges have been pointed below [5, 15].

- The development of strategies and processes to overcome the materials processing limitations in WAAM.

- The heat input from the welding torch to melt the weld material to be controlled carefully because this issue leads to many of the weld defects.

- The ease of integration with existing welding equipment and the adaptability of the process to the range of materials to be produced.

- As WAAM is purely a commercial manufacturing process for the fabrication to provide high versatility, it will be more important that the few of the additional processes are to be identified and they are capable of enhancing quality in multiple materials.

- It has been discussed that path planning in WAAM plays a main role in the successful fabrication, therefore path planning skills for the operators are to essentially be improved.

- Effects on material properties are to be kept towards increasing trend as far as possible for the WAAM fabrication in live always.

- Some of the material removal processes are likely to be developed and few are to comprise for the future successful WAAM.

- Minimizing inhomogeneous material properties attributed to geometrical changes without any compromise.
- Technical advancements to manipulate grain size and solidification mode in the fabricated part is needed for WAAM to advance.
- Simulation software to develop computational simulations regarding the numerical analysis.
- Post-processing time and cost for WAAM component are to be addressed fundamentally in order to take WAAM to next level.
- Knowledge of development machine and material properties will also help in feature addition.
- The urgency of highly capable WAAM machines that may combine manufacturing processes from a number of process classes to competently transform many raw materials into ended parts with minimal post-processing.

8. Conclusions

The concept of WAAM is now well spreading across the globe due to its wide acceptability by leading industries and crucial sectors for the fabrication of stainless steel and other metal components of interest at good cost benefits but without compromising the quality. This text clearly states the fundamentals of WAAM. The AM technique was well illustrated followed by its different classifications and techniques. The working of Wire Arc Additive Manufacturing (WAAM), a variant AM process has been written in lucid manner. Classifications of WAAM have been given with importance of familiar and capable welding processes experimented worldwide followed by synonym terms practiced across the different research group nations. The welding processes frequently used for WAAM have been well presented followed by various key welding process parameters. Moreover, the well-exercised materials of different alloy classifications have been detailed. As an example of fabrication, the details about the fabrication of metal component using GMAW based WAAM has been presented. Most common welding defects frequently occurred have been presented with the causes for easy understanding followed by some of the challenges for future to address for the successful fabrication of variety of metal components using this welding based AM. It is believed that the information furnished here would be supportive for the budding engineers and industry personnel who want to learn the AM fabrication using WAAM.

Acknowledgements

The authors would like to acknowledge the Department of Mechanical Engineering, Coimbatore Institute of Engineering and Technology for providing valuable information from their Welding Research Laboratory.

Author details

Maruthasalam Sowrirajan[1*], Selvaraj Vijayan[2] and Munusamy Arulraj[3]

1 Department of Mechanical Engineering, Coimbatore Institute of Engineering and Technology, Coimbatore, India

2 Department of Mechanical Engineering, National Institute of Technology, Tiruchirappalli, India

3 Department of Mechanical Engineering, Sri Krishna Polytechnic College, Coimbatore, India

*Address all correspondence to: sowrirajanmechanical@gmail.com

References

[1] Ziolkowski M, Dyl T. Possible applications of additive manufacturing technologies in shipbuilding: A review. Machines. 2020;**8**(4):84. DOI: 10.3390/machines8040084

[2] Zocca A, Colombo P, Gomes CM, Günster J. Additive manufacturing of ceramics: Issues, potentialities, and opportunities. Journal of the American Ceramic Society. 2015;**98**(7):1983-2001. DOI: 10.1111/jace.13700

[3] Gao W, Zhang Y, Ramanujan D, Ramani K, Chen Y, Williams CB, et al. The status, challenges, and future of additive manufacturing in engineering. Computer-Aided Design. 2015;**69**:65-89. DOI: 10.1016/j.cad.2015.04.001

[4] Lakhdar Y, Tuck C, Binner J, Terry A, Goodridge R. Additive manufacturing of advanced ceramic materials. Progress in Materials Science. 2021;**116**:100736. DOI: 10.1016/j.pmatsci.2020.100736

[5] Cunningham CR, Flynn JM, Shokrani A, Dhokia V, Newman ST. Invited review article: Strategies and processes for high quality wire arc additive manufacturing. Additive Manufacturing. 2018;**22**:672-686. DOI: 10.1016/j.addma.2018.06.020

[6] Xia C, Pan Z, Polden J, Li H, Xu Y, Chen S, et al. A review on wire arc additive manufacturing: Monitoring, control and a framework of automated system. Journal of Manufacturing Systems. 2020;**57**:31-45. DOI: 10.1016/j.jmsy.2020.08.008

[7] Sowrirajan M et al. A new approach to the fabrication of thin-walled plate component through typical wire arc additive manufacturing. Journal of Advanced Mechanical Sciences. 2022;**1**:8-13. DOI: 10.5281/zenodo.6330531

[8] Jafari D, Vaneker TH, Gibson I. Wire and arc additive manufacturing: Opportunities and challenges to control the quality and accuracy of manufactured parts. Materials & Design. 2021;**202**:109471. DOI: 10.1016/j.matdes.2021.109471

[9] Chaturvedi M, Scutelnicu E, Rusu CC, Mistodie LR, Mihailescu D, Subbiah AV. Wire arc additive manufacturing: Review on recent findings and challenges in industrial applications and materials characterization. Metals. 2021;**11**(6):939. DOI: 10.3390/met11060939

[10] Treutler K, Wesling V. The current state of research of wire arc additive manufacturing (WAAM): A review. Applied Sciences. 2021 Jan;**11**(18):8619. DOI: 10.3390/app11188619

[11] Wu B, Pan Z, Ding D, Cuiuri D, Li H, Xu J, et al. A review of the wire arc additive manufacturing of metals: Properties, defects and quality improvement. Journal of Manufacturing Processes. 2018;**35**:127-139. DOI: 10.1016/j.jmapro.2018.08.001

[12] Rosli NA, Alkahari MR, Abdollah MF, Maidin S, Ramli FR, Herawan SG. Review on effect of heat input for wire arc additive manufacturing process. Journal of Materials Research and Technology. 2021;**11**:2127-2145

[13] Kandavel N, Krishnaraj C, Dhanapal P, Sowrirajan M. Assessing the feasibility of fabrication and welding of nickel-alloyed ductile iron through the evaluation of tensile properties and mechanical characterization. Proceedings of the Institution of Mechanical Engineers, Part E: Journal of Process Mechanical Engineering. 2021. DOI: 10.1177%2F09544089211051639

[14] Bhatt PM, Kulkarni A, Kanyuck A, Malhan RK, Santos LS, Thakar S, et al. Automated process planning for

conformal wire arc additive manufacturing. The International Journal of Advanced Manufacturing Technology. 2022;**7**:1-26. DOI: 10.1007/s00170-021-08391-7

[15] Singh SR, Khanna P. Wire arc additive manufacturing (WAAM): A new process to shape engineering materials. Materials Today: Proceedings. 2021;**44**:118-128. DOI: 10.1016/j.matpr.2020.08.030

Chapter 8

Nitrogen Supersaturation of AISI316 Base Stainless Steels at 673 K and 623 K for Hardening and Microstructure Control

Tatsuhiko Aizawa, Tomomi Shiratori, Tomoaki Yoshino, Yohei Suzuki and Takafumi Komatsu

Abstract

The high-density plasma nitriding at 673 K and 623 K was employed to make 10% of nitrogen supersaturation on AISI316 base austenitic stainless steels. The processing parameters and nitrogen-hydrogen gas flow ratio were optimized to increase the yield of N2+ ion and NH-radical for efficient nitriding. The nitrided AISI316 specimens were prepared for multidimensional analysis to describe the fundamental features of low-temperature plasma nitriding. First, macroscopic evaluation revealed that nitrogen supersaturation induced the γ-lattice expansion and the higher nitrogen content than 4% of mass in depth. The mesoscopic analysis describes the holding temperature and initial grain-size effects on the microstructure changes. Plastic straining, grain-size refinement, and nitrogen zone-boundary diffusion processes advance with nitrogen supersaturation to drive the inner nitriding behavior. The microscopic analysis explains the microstructure refinement, the two-phase structuring, and the microstructure modification. Through this multidimensional analysis, the essential characteristics of the low-temperature plasma nitriding of 316 austenitic stainless steels were precisely understood to extend the engineering treatise on the bulk nitrogen stainless steels for surface modification and treatment of stainless steels by nitriding. This plasma nitriding was applied to strengthen and harden the AISI316 wire surfaces toward its application on surgery wires.

Keywords: high-density low-temperature plasma nitriding, AISI316 stainless steels, multidimensional analysis, nitrogen supersaturation, plastic straining, microstructure refinement, two-phase structuring, nitrogen zone-boundary diffusion, surgery wires

1. Introduction

Stainless steel was invented in 1915 as non-rusting steel with high chromium and nickel contents by the alloying design [1]. Since then, various kinds of stainless steels have been developed to improve their features in suitable to each application; e.g., austenitic, martensitic, ferritic, and two-phase stainless steels are widely utilized in the present society [2]. As one of the high-strength austenitic stainless steels, AISI316 plates, bars, and wires have been widely utilized in industries [3]. In addition, it has a family of industrial grades such as AISI316L, low carbon AISI316, to improve the corrosion toughness [4] and AISI316LN, nitrogen-bearing AISI316, to improve the erosion and wear toughness [5]. In the history to develop these austenitic stainless steels, nitrogen was highlighted as its effective alloying element to reduce the amount of nickel consumption in fabrication. In this development of HNSS (High Nitrogen Stainless Steels), the role of nitrogen solute contents in the mechanical properties of stainless steels has been studied in [6]. In parallel to these studies on HNSS, various processes were developed to make nitrogen supersaturation to the γ-phase steels beyond their maximum nitrogen solubility. However, the nitrogen content in HNSS is still limited by 1 mass%. An increase of the nitrogen solute content in HNSS or nitrogen supersaturation is still a challenge to significantly improve the mechanical and functional properties of HNSS.

In parallel to R & D on HNSS, the nitriding process is another route to utilize this nitrogen in the surface treatment and modification of Fe-Cr base alloys and stainless steels [7]. The gas and liquid nitriding processes were first employed to form the thick nitrided layer with the use of the ammonia gas, the chloride ion, and cyanic liquids. The plasma nitriding process was gradually selected as an environmentally benign route of surface treatment instead of those processes [7, 8]. This plasma nitriding process is classified into two categories on the dependence of holding temperature and duration [8]; e.g., high-temperature plasma nitriding (HT-PN) with nitride precipitation into the nitrided layer for hardening, and low-temperature plasma nitriding (LT-PN) with the nitrogen supersaturation into the depth of matrix for hardening, strengthening and improvement of wear−/corrosion toughness.

Figure 1 depicts this categorizing on the plasma nitriding of AISI316 stainless steels. Above the master curve, the CrN (Chromium Nitride) or iron nitride precipitation governs the hardening process by HT-PN of AISI316 steels. Those nitrides precipitate in the AISI316 matrix to strengthen the stainless steel; their surface hardness increases but their nitrided layer thickness decreases with increasing the chromium content [7]. The nitriding process is governed by the nitrogen body-diffusion mechanism, so that the nitrogen solute content exponentially decreases from the surface to the depth [9], the nitrogen content at the surface is limited by the maximum solubility of 0.3 mass% [7], and the square of nitrided layer thickness is proportional to the holding duration [7, 10]. The metal chromium content decreases by synthesis of CrN in the nitrided layer and results in significant loss of corrosion toughness, intrinsic to AISI316. On the other hand, no nitrides are synthesized in the nitrided layer by LT-PN; nitrogen solute supersaturates the AISI316 matrix to form a thicker nitrided layer than 50 μm, to harden this layer up to 1400 HV, and to modify the microstructure of matrix at 673 K for 14.4 ks (or 4 h) [11–15]. Remember in HT-PN that 1) thinner nitrided layer is only formed in case of the high chromium contents, 2) surface hardness is limited by 1200 HV, 3) no significant change of microstructure is observed in the nitrided layer, and 4) microstructure below NFE remains the same as the original AISI316 before nitriding. In particular, the LT-PN at 673 K is characterized by the nitrogen supersaturation into the austenitic stainless steels with higher nitrogen content than 4 mass% and without iron and chromium nitride precipitates [13–17]. As studied in [12, 13], this nitrided layer improved the corrosion toughness of original AISI316 stainless steels. No chromium content was reduced even after the nitriding process due to the nitrogen supersaturation. Hence, LT-PN becomes a candidate processing to modify the AISI316 product surfaces to a HNSS layer with high nitrogen solute content and to improve its mechanical and functional properties.

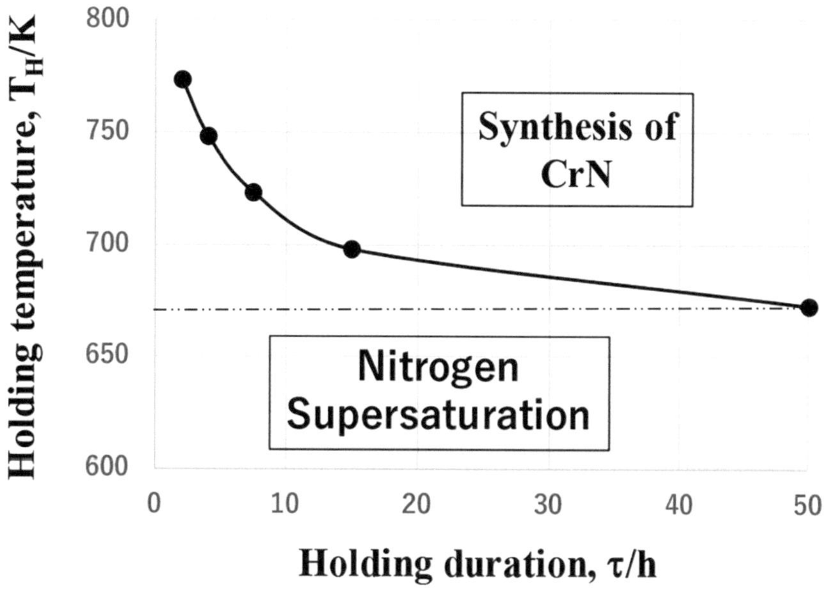

Figure 1.
Two categories in the plasma nitriding process on the dependence on the holding temperature and duration.

In the present paper, the high-density plasma nitriding process is redesigned by using the plasma diagnosis to be working under the optimum conditions. In particular, OES (Optical Emissive-light Spectroscopy) is employed to search for the most suitable nitrogen – hydrogen gas flow rate to attain the higher yield of N_2^+ ions and NH radicals for efficient nitriding. AISI316 specimen is nitrided respectively at 673 K and 623 K under the optimized conditions to describe the nitrogen supersaturation process by using the multidimensional analysis. Macroscopic evaluation on the formation of the nitrided layer is performed by XRD and SEM–EDX. EBSD is employed to make a mesoscopic evaluation of the holding temperature and initial grain-size effects on the nitriding behavior in the nitrided layer and below the NFE. The plastic straining, the microstructure refining, and the two-phase structuring advance in synergy with nitrogen supersaturation and diffusion through the nitrided layer. This synergic relation among these processes is common to every nitriding behavior in LT-PN, irrespective of the holding temperature and initial granular structure. Microscopic analysis with the use of STEM is employed to describe the microstructure refining and two-phase structuring in the nitrided AISI316 at 673 K. STEM analysis directly prove that the microstructure refining process is driven by the shear localization in the plastic straining, two-phase structuring is induced by the microscopic disturbance of nitrogen content with its different chemical compatibility to iron, nickel, and chromium in AISI316, and that plastic straining still modifies the crystalline structure at the absence of nitrogen solutes even below NFE. The above multidimensional analysis demonstrates that low-temperature plasma nitriding is driven by the synergic relation of plastic straining, microstructure refining, and two-phase structuring with the nitrogen supersaturation and zone-boundary diffusion processes. The homogeneous inner nitriding of initially fine-grained AISI 316 is self-sustained by this synergic effect not to ignite the localization in nitrogen supersaturation even at 623 K. This self-sustainable nitriding is attractive to make surface treatment of stainless steel medical parts. Especially, the nitrided AISI316 wire is straightforwardly utilized as a surgery wire by its uniform high surface hardness and high loading capacity.

2. High-density plasma nitriding system

The high-density RF (Radio Frequency) - DC (Direct Current) plasma nitriding system is depicted in **Figure 2a**. It consists of the vacuum chamber, the evacuation system, the gas supply, the RF/DC-power generator, and the control panel. On the basis of the hollow cathode device [13, 18], the experimental setup was redesigned with the use of plasma diagnosis [19]. **Figure 2b** depicts a typical hollow-cathode setup for the present nitriding processes. High brightness only in the inside of the hollow proves how much the ion and radical populations are present in the ignited plasmas.

Figure 2.
High-density plasma nitriding system. a) Overview of RF-DC plasma nitriding system, and b) hollow cathode set up for high-density nitriding process.

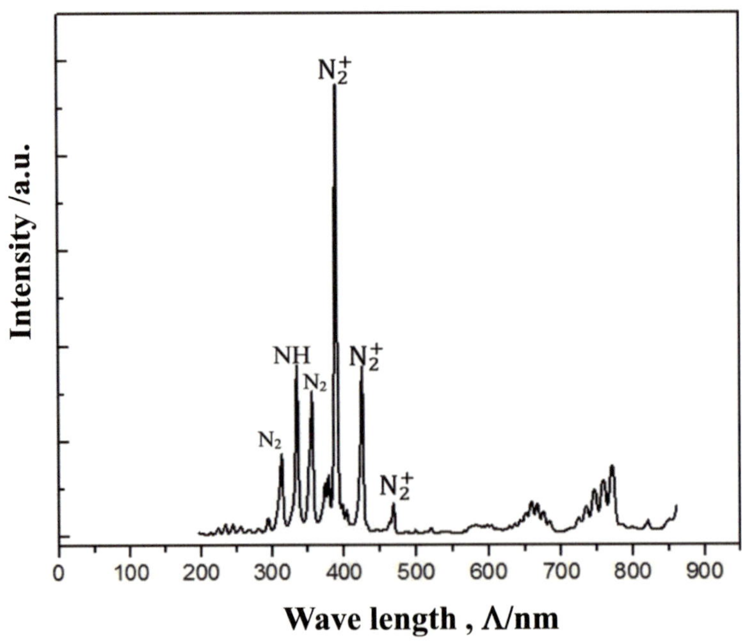

Figure 3.
A typical optical emission spectrum from the nitrogen and hydrogen plasmas is measured by the plasma diagnosis.

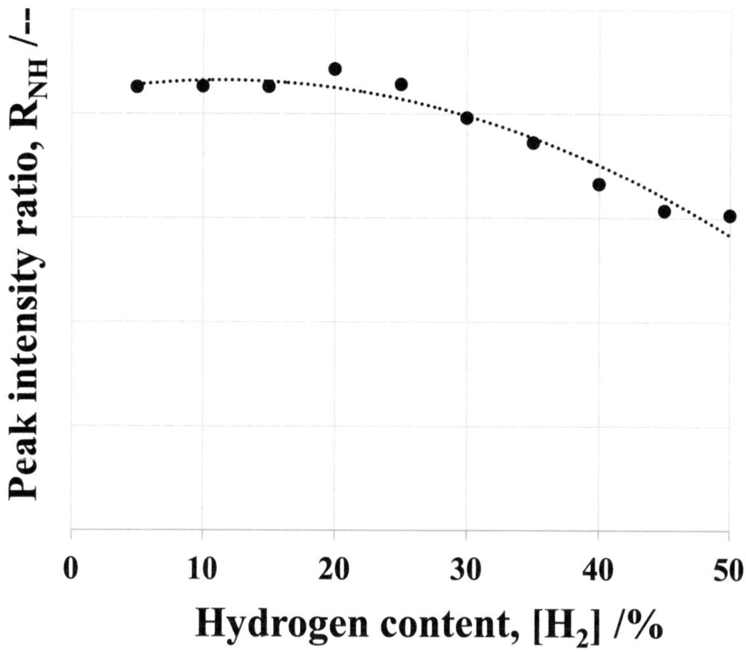

Figure 4.
Variation of the intensity ratio (R_{NH}) of NH-radical peak to N_2^+ peak by OES with increasing the hydrogen gas content in N_2 + H_2 mixture gas.

Among various plasma diagnosis methods, OES (Optical Emissive-light Spectroscopy) was employed to describe the activated species in the nitrogen – hydrogen plasmas. **Figure 3** depicts a typical spectrum. Both N_2^+ and NH are main species with high intensity. After [20], N_2^+ is a mother species to generate a series of nitrogen ions by electron attachment and detachment reactions in plasmas. The N_2^+ peak with the highest intensity and the NH peak at Λ = 340 nm was employed for diagnosis.

After optimizing the relationship among the RF-voltage, the DC-bias, and the pressure, the nitrogen–hydrogen gas ratio was varied to search for optimization of the intensity ratio (R_{NH}) of NH-radical peak to N_2^+ peak. **Figure 4** shows the variation of R_{NH} with increasing the hydrogen gas content $[H_2]$ to the nitrogen–hydrogen mixture gas. When $[H_2]$ < 5%, less NH-peaks were detected by OES; R_{NH} increases with increasing $[H_2]$. After R_{NH} becomes maximum at $[H_2]$ = 20-25%, this ratio significantly decreases with increasing $[H_2]$ due to the hydrogen quenching effect in the plasma synthesis of NH radicals. In the following nitriding experiments, the nitrogen and hydrogen gas flow rate were respectively fixed by 160 mL/min and 30 mL/min after this plasma diagnosis.

3. Macroscopic characterization on the plasma nitrided AISI316

XRD and SEM–EDX were utilized to describe the nitrogen supersaturation and nitrided layer formation in LT-PN of the AISI316 at 673 K and 623 K or 14.4 ks. **Figure 5** compares the three XRD diagrams of un-nitrided and nitrided AISI316 specimens.

The bare AISI316 is characterized by γ (111) peak at 2θ = 44.3° and γ (200) at 2θ = 51.7°, respectively. These peaks shift to γ_N peaks in the shallow 2θ directions in both nitrided AISI316 specimens at 673 K and 623 K. This peak shift proves that

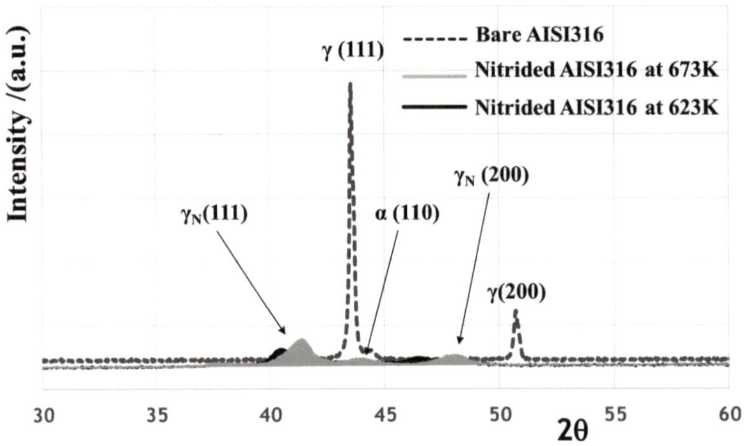

Figure 5.
XRD diagrams for plasma nitrided AISI316 specimens at 673 K and 623 K for 14.4 ks.

Nitrided AISI316	2θ-shift for γ (111) (degree)	Lattice strain (%)	2θ-shift for γ (200) (degree)	Lattice Strain (%)
673 K	44.3° → 41.4°	6.78%	51.7° → 48.05°	7.06%
623 K	44.3° → 40.5°	9.04%	51.7° → 46.6°	10.17%

Table 1.
Estimate the induced lattice strain by the nitrogen supersaturation into the nitrided AISI316 specimen at 673 K and 623 K.

nitrogen interstitial occupies the vacancy sites of austenitic lattice in AISI316 and the lattice expands by itself. That is, the nitrogen supersaturation to AISI316 is first defined by this lattice expansion in **Figure 5**. After theoretical study in [21], this lattice expansion is induced by the occupation of interstitial nitrogen atoms to the octahedral vacancy sites. Let us estimate the lattice strains by this nitrogen supersaturation. **Table 1** summarizes the peak shift and lattice strain induced by the

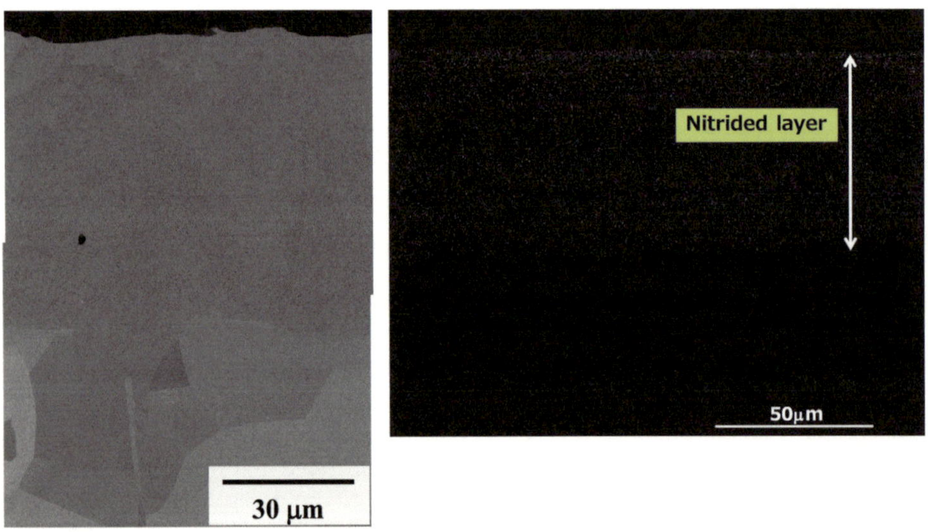

Figure 6.
SEM image and EDX mapping on the cross-section of nitrided AISI316 specimen at 673 K for 14.4 ks.

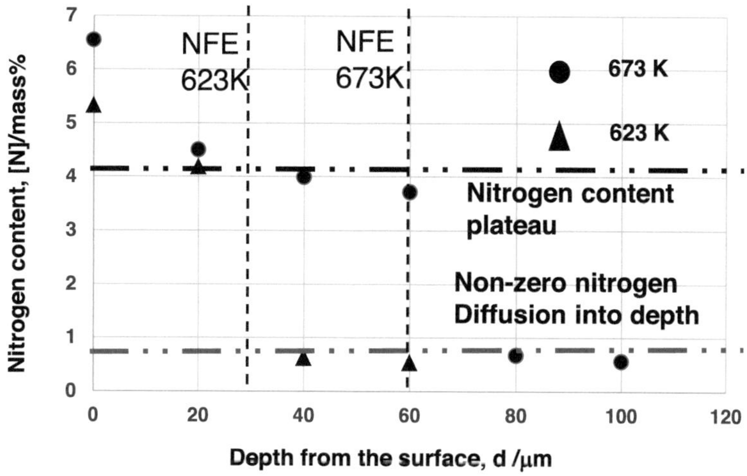

Figure 7.
Nitrogen solute content depth profiles for the nitrided AISI316 at 673 K and 623 K for 14.4 ks.

Nitrided AISI316	Nitrided layer thickness (μm)	Maximum surface nitrogen content (mass%)	Plateau nitrogen content (mass%)
673 K	60	6.5	4.0
623 K	30	5.3	4.2

Table 2.
Nitrogen content analysis on the nitrided layer of AISI316 at 623 K and 673 K for 14.4 ks.

nitriding at 673 K and 623 K, respectively. The nitrogen supersaturation induces the lattice strain of 7% at 673 K and 10% at 623 K in the nitrided layer, respectively.

SEM–EDX was utilized to describe the nitrogen solute distribution in the depth of nitrided AISI316 at 673 K and 623 K, respectively. SEM image and nitrogen mapping on its cross-section in **Figure 6** show that a thick nitrided layer with a thickness of 60 μm is formed to have uniform nitrogen solute content. To be noticed, the microstructure below NFE is slightly modified by the nitriding. The nitrogen content depth profiles are also depicted in **Figure 7**. Irrespective of the holding temperature, these profiles have a plateau with the constant nitrogen content of 4 mass% till NFE, as listed in **Table 2**.

This nitrogen content decreases at the vicinity of NFE, but a significant nitrogen content of 0.5 mass% is present even below NFE. The role of non-zero nitrogen content blow NFE is considered in the mesoscopic evaluation on the inner nitriding behavior.

4. Mesoscopic characterization on the plasma nitrided AISI316

EBSD was employed to describe the microstructure change before and after nitriding and to analyze the effects of holding temperature and initial grain size on nitriding behavior. In the EBSD analysis, three items are employed to disclose the mesoscopic view on the nitriding behavior; e.g., the phase mapping, the KAM (Kernel Angle Misorientation) distribution, and the IPF (Inverse Pole Figure) profile. As shown in **Figure 5**, the nitrided AISI316 is composed of the nitrogen supersaturated γ-phase (γ_N) and α-phase (α_N). The phase mapping on the cross-section of the nitrided layer depicts that nitrogen supersaturation induces the lattice expansion and transforms γ- to α-zones and that plastic straining occurs in the nearest neighboring γ-zones to each lattice expanding zones. The dislocations are generated to compensate for the mismatched strains between the unsaturated γ-phase zones and the lattice strained α−/γ-zones [22, 23].

KAM profile represents the equivalent plastic strain distribution which is induced by the lattice expansion during the nitrogen supersaturation after [24]. IPF provides information on the grain refinement and subgrain formation with crystallographic spin-rotation.

Figure 8 depicts the phase mapping, the KAM distribution, and IPF profile on the cross-section of nitrided AISI316 at 623 K. As had been discussed in [22, 23, 25], most of the microstructure above NFE at the depth of 30 μm from the surface has two-phase structure, high plastic strains, and fine grain structure. Due to relatively homogeneous nitrogen supersaturation, the plastic straining and microstructure refinement processes co-work with the nitrogen supersaturation and zone-boundary diffusion processes. However, this synergic nitriding process commences to localize by itself near NFE and completely localizes below NFE. Since the α-phase is continuously formed along the a-path, the nitrogen atoms diffuse along this a-path to the depth of AISI316 matrix. This zone-boundary diffusion assists the further nitrogen diffusion into the neighboring grains such as A- and B-grains. As seen in **Figure 8**, the phase transformation, the plastic straining, and the grain size refinement advance even in A- and B-grains with this nitrogen supersaturation and diffusion processes.

Let us describe this local nitrogen supersaturation into A- and B-grains. **Figure 9** depicts the phase mapping, the strain distribution, and the grain refining in these grains. Compared between **Figure 9a** and **b**, the α-phase zones in A-grain are formed in the absence of plastic strains, and, the high KAM zones correspond to the γ-phase. Since these plastic strains are induced by the lattice expansion in the nitrogen supersaturated γ-phase, these lines and zones with high KAM

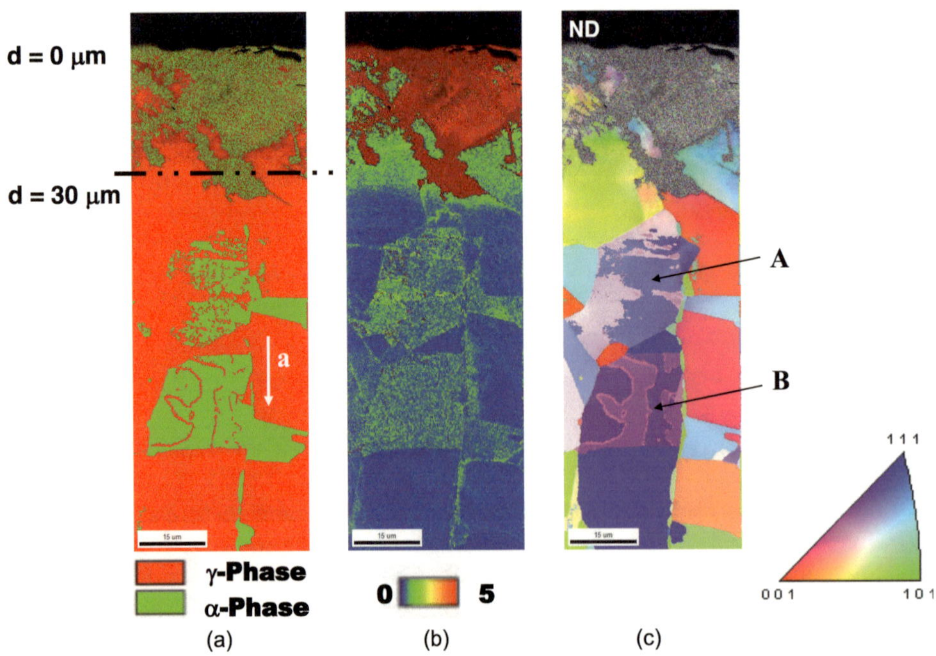

Figure 8.
EBSD analysis on the cross-section of the nitrided AISI316 at 623 K for 14.4 ks. a) Phase mapping, b) KAM distribution, and c) inverse pole figure in the ND direction.

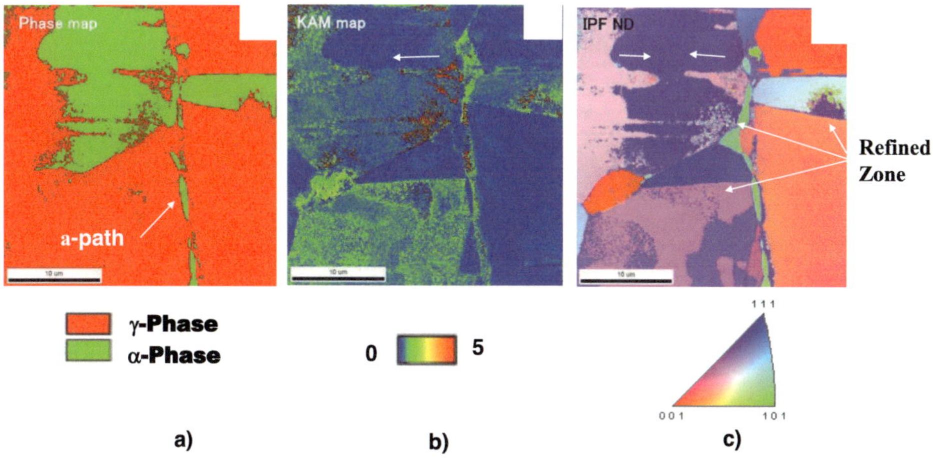

Figure 9.
EBSD analysis on the cross-section below NFD for the nitrided AISI316 at 623 K for 14.4 ks. a) Phase mapping, b) KAM distribution, and c) inverse pole figure in the ND direction.

correspond to the nitrogen diffusing paths into A-grain. Most of these diffusing paths terminate intermediately so that nitrogen supersaturation and microstructure refinement processes also stop in the inside of A-grain. Comparing **Figure 9b** and **c**, high KAM zones in **Figure 9b** is just corresponding to fine-grained zones in **Figure 9c**. This demonstrates that the grain-size refinement is induced by the plastic straining.

Figure 9 reveals that the nitrogen supersaturation and zone-boundary diffusion processes localize below NFE. These processes co-work with the plastic straining, the two-phase structuring, and microstructure refinement in local to modify the crystallographic structure far below NFE. This heterogeneous nitrogen supersaturation process must be changed to be more homogeneous by controlling the external and internal nitriding conditions.

First, the holding temperature is increased from 623 K to 673 K to investigate its effect on this mode change. **Figure 10** depicts the phase mapping, the plastic strain distribution, and the IPF profile on the cross0section of nitrided AIS316 at 673 K for 14.4 ks.

Although the a-path for zone-boundary nitrogen diffusion is formed below NFE together with plastic strains and refined microstructures, the homogeneous nitrogen supersaturation process advances above NFE. The two-phase structuring in **Figure 10a**, the plastic straining in **Figure 10b**, and microstructure refining in **Figure 10c** co-work in synergy and co-terminates at d = 60 μm. The heterogeneous nitrogen supersaturation is suppressed to the local area below NFE. This proves that the mode-change from the heterogeneous nitriding to the homogeneous one is performed by simply increasing the holding temperature.

In addition to this external item, the initial grain size refinement is selected as an internal item to make the mode change. The intense rolling was employed to reduce the original AISI316 plate thickness by 90%. **Figure 11** shows the phase mapping, the plastic strain distribution, and the IPF profile on the surface of the fine-grained AISI316 (or GF-AISI316) plate. As well known in the rolling and stamping of AISI304 plates [26] and AISI316 bars [27], the transformation from γ-phase to α-phase is induced into AISI316 by this intense rolling as shown in **Figure 11a**. The agreement between the α-phase zones and the high strained zones proves this strain-induced transformation in comparison to **Figure 11a** and **b**. The initial grain size with its average of 15 μm is reduced to 1.5 μm in **Figure 11c** by rolling.

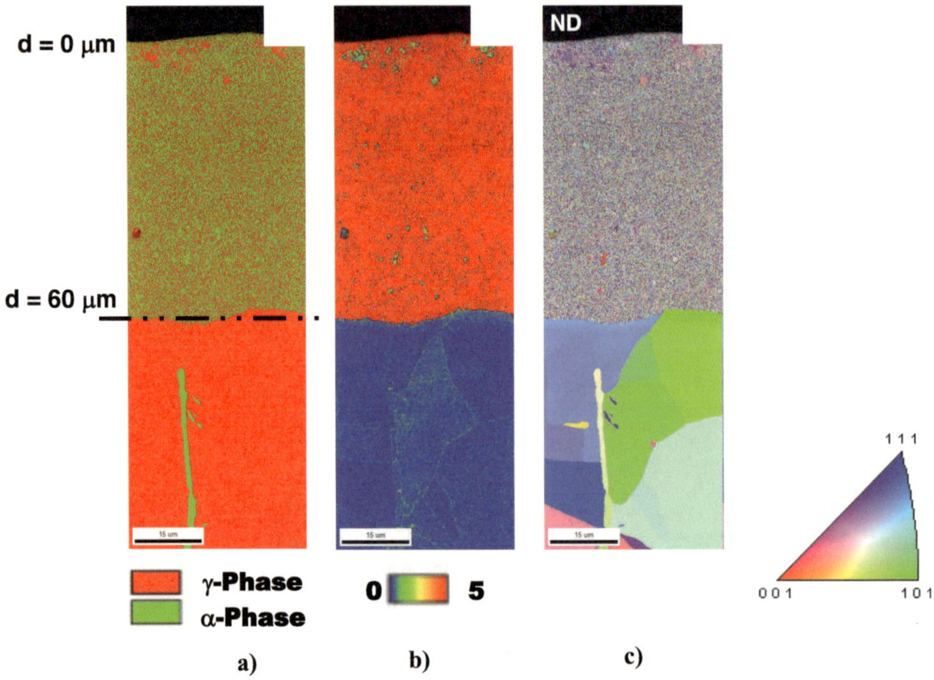

Figure 10.
EBSD analysis on the cross-section of the nitrided AISI316 at 673 K for 14.4 ks. a) Phase mapping, b) KAM distribution, and c) inverse pole figure in the ND direction.

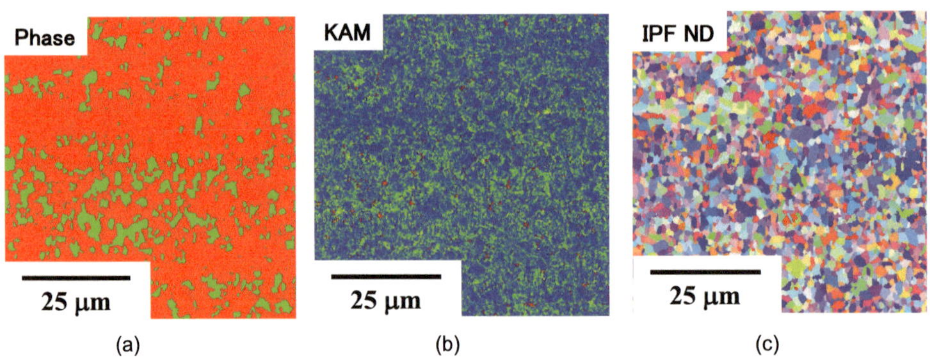

Figure 11.
EBSD analysis on the surface of rolled AISI316 plate specimen before nitriding. a) Phase mapping, b) KAM distribution, and c) inverse pole figure in ND direction.

This GF-AISI316 specimen was nitrided at 623 K for 14.4 ks to investigate the effect of initial grain size on the mode change in nitrogen supersaturation. **Figure 12** shows the SEM image and nitrogen mapping on the cross-section of nitrided GF-AISI316 at 623 K for 14.4 ks. The uniform nitrided layer with a thickness of 40 μm was formed with a fine microstructure above NFE. The microstructure below NFE is also homogeneous and looks to be the same as the microstructure before nitriding in **Figure 12**. The hardness depth profile and the nitrogen solute content depth profile across NFE are respectively depicted in **Figure 13**.

The hardness with its average of 1400 HV above NFE drastically decreases down to the matrix hardness of 250 HV just across NFE in **Figure 13a**. The nitrogen solute

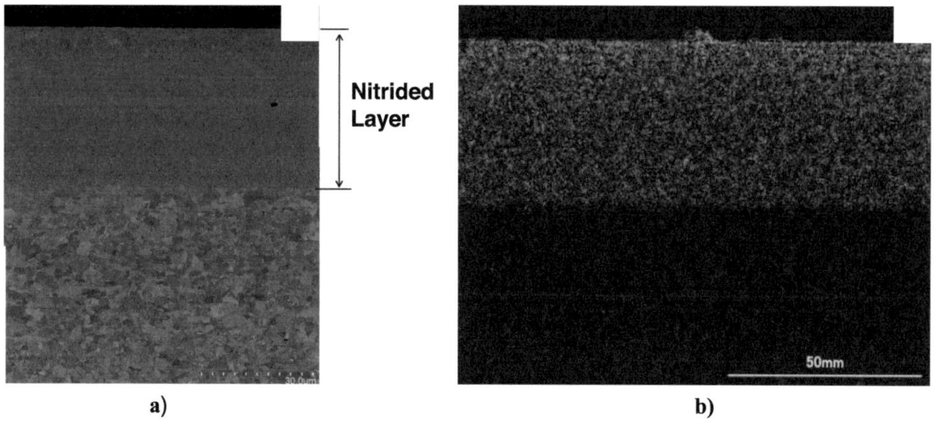

Figure 12.
SEM and nitrogen mapping on the cross-section of the rolled AISI316 specimen after nitriding at 623 K for 14.4 ks. a) SEM image on the cross-section, and b) nitrogen mapping to the depth.

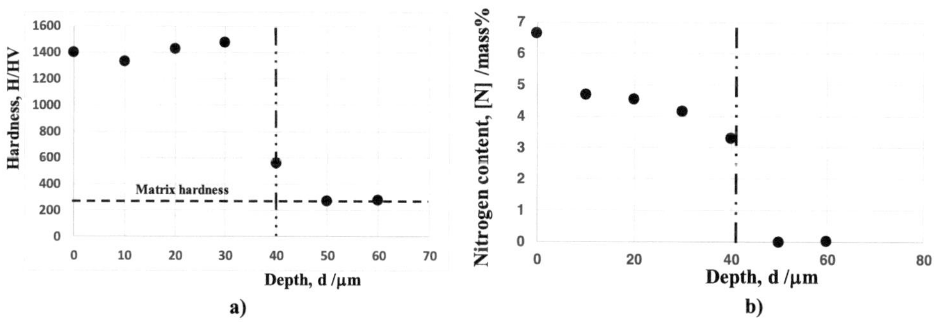

Figure 13.
Hardness and nitrogen content depth profiles of the rolled AISI316 specimen after nitriding at 623 K for 14.4 ks. a) Hardness depth profile, and b) nitrogen content depth profile

content depth profile with the average of 4 to 5 mass% also decreases down to zero across NFE in **Figure 13b**. These profiles reveal that the homogeneous nitrogen supersaturation advances to NFE and terminates at NFE. This is completely different from the heterogeneous nitrogen supersaturation process across NFE when using the AISI316 with the average grain size of 15 μm. Let us make EBSD analysis on the cross-section of nitrided FG-AISI316 with a comparison between **Figures 8** and **14**. As shown in **Figure 14a**, the two-phase structuring, the plastic straining, and the grain size refining take place only above NFE and no processes advance across NFE. The two-phase, highly strained, and grain-refined zone in **Figure 14a–c** is just equivalent to the nitrided layer. The phase map, the KAM distribution, and IPF profile below NFE in **Figure 14** are the same as those in **Figure 11**. That is, no nitrogen supersaturation takes place below NFE.

This comparison of EBSD results among **Figures 8, 11**, and **14** reveals that the initial grain size refinement has a significant influence on the synergic relationship among the nitrogen supersaturation, the nitrogen zone-boundary diffusion, the plastic straining, the two-phase structuring, and the microstructure refining.

Let us reconsider the heterogeneous and homogeneous nitrogen supersaturation processes and their mode change. As depicted in **Figure 8**, the heterogeneous nitrogen supersaturation gradually turns to be homogeneous according to the

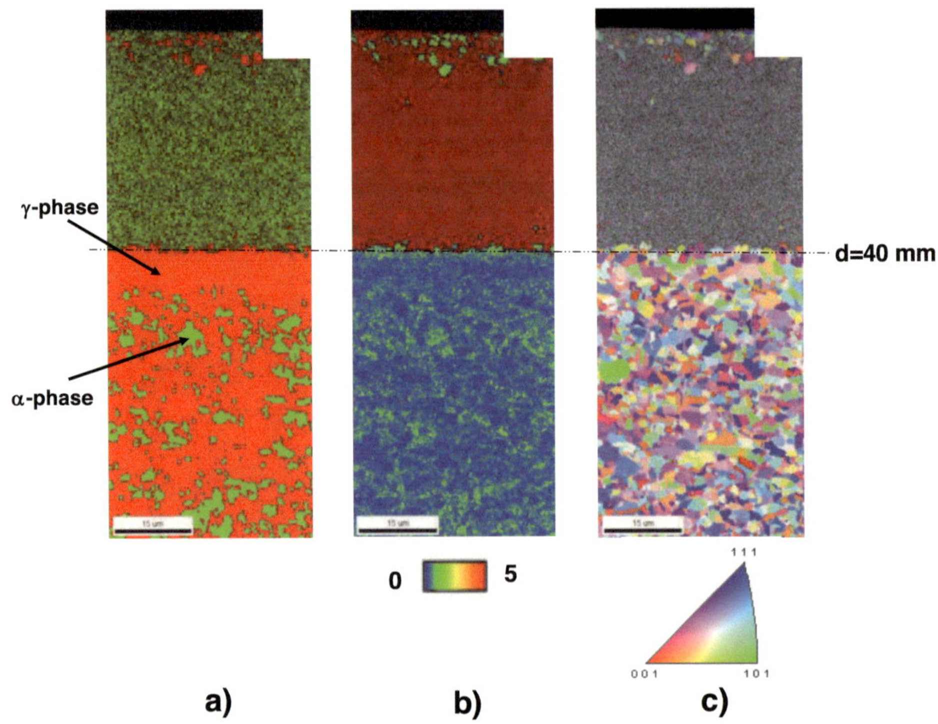

Figure 14.
EBSD analysis on the surface of rolled AISI316 plate specimen after nitriding at 673 K for 14.4 ks. a) Phase mapping, b) KAM distribution, and c) inverse pole figure in ND direction.

nitrogen boundary diffusion mechanism change from the localized path to the network. Under the localized boundary diffusion mechanism, the nitrogen supersaturation, the plastic straining, the grain refining, and the phase transformation only advance at the vicinity of zone boundaries. Neglecting this localization, the synergic relationship to drive the inner nitriding process by heterogeneous nitrogen supersaturation is the same as that by homogeneous nitrogen supersaturation. Hence, if the localized nitrogen boundary diffusion is revised by the zone-boundary diffusion in a network, this heterogeneous nitrogen supersaturation process could be controlled to change itself to a homogeneous process. Both the holding temperature increase and the initial grain size refinement work as an external and internal trigger to enhance the nitrogen zone boundary diffusion mechanism.

In the case of the higher holding temperature, the nitrogen boundary diffusion rate is enhanced in similar manner to the increase of body diffusion rate in the traditional plasma nitriding processes such as ion- and radical-plasma nitriding. Although those HT-PN processes required a higher temperature than 773 K enough to sustain the diffusion path, the holding temperature of 673 K is enough to drive the homogeneous nitriding.

The HT-PN processes have no initial grain-size refinement effect on their nitriding behavior. On the other hand, the heterogeneous nitriding even at 623 K completely changes to be homogeneous when using the fine-grained AISI316 substrates. This mode change proves that grain-boundary diffusion works as a network to supply a sufficient amount of nitrogen solutes enough to sustain the synergic nitriding process at every spot above NFE. As demonstrated in **Figures 12–14**, NFE works just as a front end for homogeneous nitrogen supersaturation and diffusion.

5. Microscopic characterization on the plasma nitrided AISI316

Through the macroscopic and mesoscopic analysis, the plasma nitrided AISI316 is characterized by the refined and two-phase microstructure, by the plastically strained microstructure, and by the modified microstructure below NFE. STEM was employed to analyze the microstructure to microscopically describe the grain-size refining, the two-phase structuring, the plastic straining, and the microstructure modification below NFE during the plasma nitriding of AISI316 at 673 K for 14.4 ks.

The Cs-corrected STEM (Spherical Aberration Corrected Scanning Transmission Electron Microscope; JEM-ARM200F; JOEL, Tokyo, Japan) was employed for this microscopic characterization on the nitrided layer. This system has a cold FEG (Field Emission Gun) with a capacity of 200 kV and a resolution of 0.08 nm by using a Cs-corrector. Dual SDD (Solid State Detector)-EDS detector was utilized for local element mapping with Cs-STEM. Argon ion milling was used to make a thin slice of the plasma nitrided AISI316 specimen for Cs-STEM analysis.

Figure 15a depicts the STEM image in the vicinity of nitrided layer surface. This surface region is composed of two zones; e.g., a smooth zone and a rough zone. The electric diffraction at the former zone is shown in **Figure 15b**. Since a single spot is only detected, this former zone has a single-crystal like microstructure. On the other hand, the electric diffraction at the latter zone consists of two or three spots as shown in **Figure 15c**. That is, this zone has a polycrystalline microstructure.

Figure 15 proves that the nitrided layer with high nitrogen solute content is composed of single-crystal and poly-crystal grains. After [28], a single crystal grain with its size less than 10 nm has no dislocations or no defects in its lattice structure. Those defects in the fine single crystals have high free energy enough to be pushed out of the inside of crystals to fine-grain boundaries. Hence, this ultra-fine single-crystal zone also has no nitrogen solute to occupy the octahedral vacancy sites in the single crystal. That is, these fine single crystals are generated by the refining process

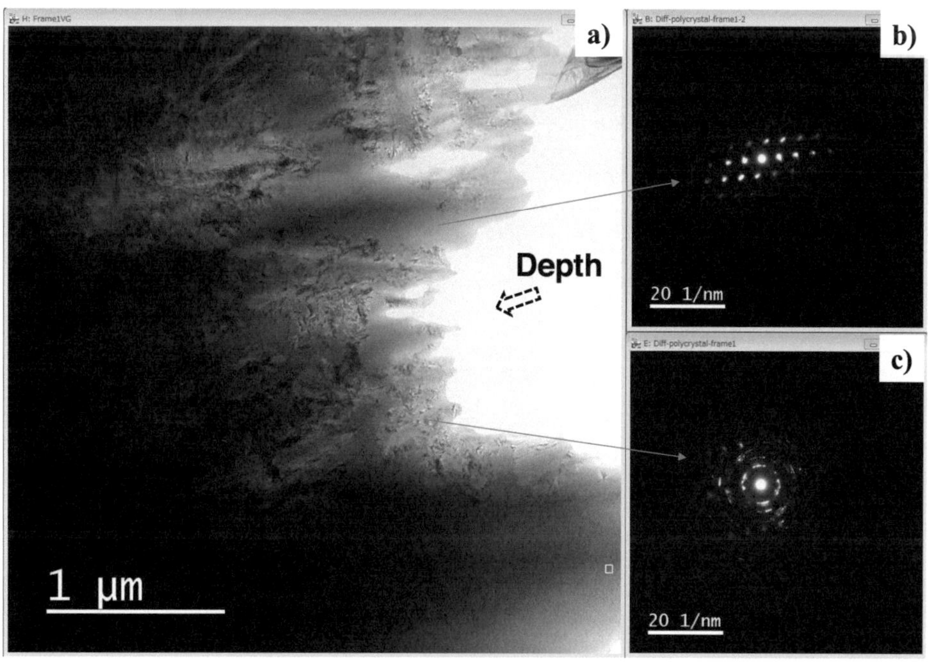

Figure 15.
TEM image on the surface of the nitrided AISI316 at 673 K for 14.4 ks. a) Cross-sectional TEM image with two zones, b) single-crystal like zone, and c) poly-crystal like zone.

through the cascading reduction of zone size, co-working with the intense plastic straining in the synergic relationship.

In later, the microstructure of this single crystal zone is precisely analyzed by HAADF (High-Angle Annular Dark-Field)-imaging, ABF (annular bright-field)-imaging, and LAADF (Low-angle annular dark-field)-imaging, respectively. Most of the microstructure in **Figure 15a** consists of the poly-crystal zones. Remember that the nitrided layer consists of the fine two-phase structures in the mesoscopic characterization. This poly-crystal zone is expected to be formed by two neighboring crystals with different lattice structures. In later, HAADF, ABF, and LAADF imaging methods are also utilized to describe this correlation.

In the STEM analysis, HAADF-imaging produces an annular dark-field image formed by very high angle, incoherently scattered electrons (Rutherford scattered from the nucleus of the atoms). ABF provides a robust technique for simultaneous imaging of light and heavy elements since its contrast has a low scaling rate with the atomic number. LAADF receives the diffracted or inelastically scattered electrons at low to medium angles (25 to 60 mrad) using an ADF (annular dark-field) detector.

Figure 16 shows the HAADF, ABF, and LADDF-images at the single-crystal zone in **Figure 14b**. ABF-imaging explains that every constituent atom is aligned in (111) direction. From LAADF imaging, this single crystal with (111) and (200) crystallographic orientations. The above nano-structural analysis reveals that this single-crystal has γ-phase, the crystallographic orientation of which coincides with the easiest slipping plane orientation of (111). Remember that the mesoscopic analysis on the nitrided layer. The γ-phase zones coexist with α-phase zones to form the fine two-phase nanostructures and the nitrogen unsaturated zones are plastically strained to compensate for the strain incompatibility between the nitrogen saturated and unsaturated zones. ABF-imaging in the above proves that a refined

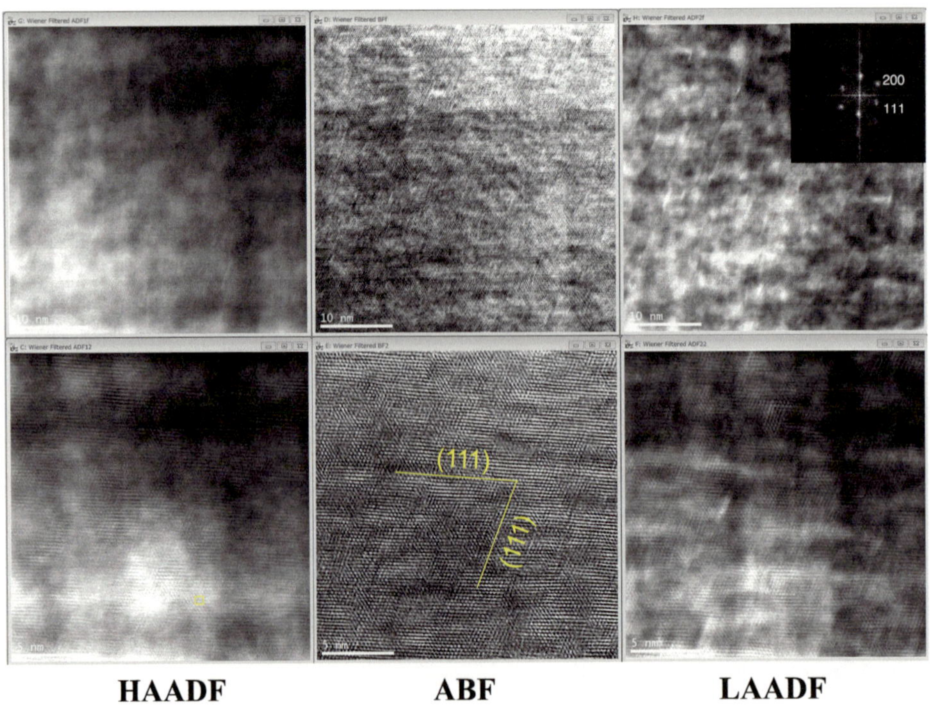

Figure 16.
HAADF, ABF, and LAADF analysis on the single-crystal-like zone at the surface of nitrided AISI316 at 673 K for 14.4 ks.

γ-phase zone with its average size of 5 nm is formed by plastically straining the nitrogen unsaturated γ-phase region.

In the mesoscopic characterization, the network of zone boundaries plays an essential role to transport the nitrogen solute from the surface to the NFE of the nitrided layer and to drive the homogeneous nitrogen supersaturation. **Figure 16** also proves that zone boundaries in this single-crystal should work as a fine network of nitrogen diffusion paths and propel the nitrogen supersaturation to the un-nitrided regions. The grain size refinement co-works with the homogeneous nitrogen supersaturation and terminates itself at this formation of ultrafine γ-phase single crystals without nitrogen solutes.

In **Figure 15**, this refined γ-phase single-crystal is neighboring to the poly-crystal zones; the other phase in the refined two-phase microstructure corresponds to this poly-crystal. Let us make STEM analysis on this point.

Figure 17 shows the HAADF, ABF, and LAADF-images on the poly-crystal zone at the vicinity of nitrided AISI316 specimen surface at 673 K for 14.4 ks. These imaging methods in low magnification prove that two zones with different crystallographic structures are aligned in series and in parallel in the inside of the poly-crystal. After STEM analysis in high resolution and dual SDD-EDX detection, either of these two zones mainly consists of nitrogen-enriched chromium or Cr (N) rich lattices. On the other hand, another zone consists of nitrogen-poor iron and nickel or Fe/Ni (N). HAADF and KAADF images also prove that two neighboring zones have different nitrogen content. In correspondence to the mesoscopic analysis on the two-phase structure, this nitrogen-rich Cr (N) has a γ_N-phase structure while the nitrogen-poor Fe/Ni(N) has α_N-phase structure.

To be described later, the microstructure of AISI316 below NFE has no separation among iron, nickel, and chromium contents. This local segregation of chromium from iron and nickel is induced by the difference in chemical compatibility to

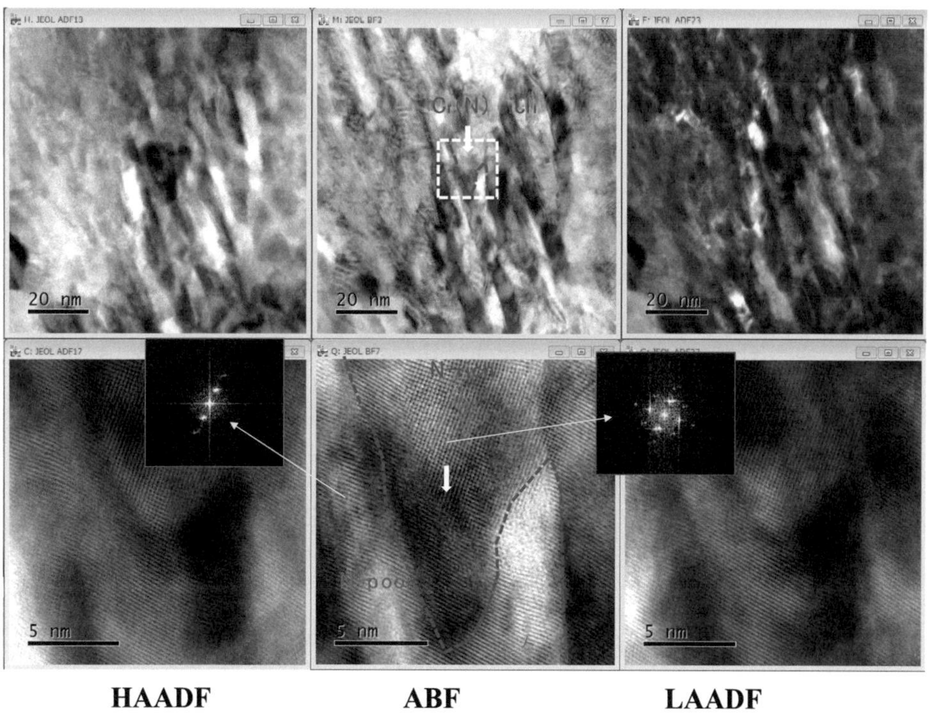

Figure 17.
HAADF, ABF, and LAADF analysis on the poly-crystal like zone at the surface of nitrided AISI316 at 673 K for 14.4 ks.

nitrogen among the three elements. In the austenitic stainless steels, chromium is a substitutional element to occupy the iron site without significant change of the original lattice constant for iron [7]. Nickel works to stabilize the fcc-structure [8]. In the absence of nitrogen solute in AISI316 substrate, its crystalline system has γ-phase while this local system has a fine mixture of γ_N- and α_N-substructures by their chemical compatibility to nitrogen solutes. This local separation of a constituent element in AISI316 among chromium, nickel, and iron, drives the nitrogen super-saturation and zone-boundary diffusion to advance into the original matrix. The zone boundaries between Cr (N) and Fe/Ni (N) work as a local network of nitrogen diffusion paths to modify the original, homogeneous AISI316 matrix to a fine two-phase crystalline structure of Cr (N) and Fe/Ni (N). This formation of nitrogen supersaturated γ- and α-phase zones reveals that nitrogen solute diffuses through the zone boundary network to the inside of each nano-crystalline zone.

Figure 17 also proves that the chemical compatibility of AISI316 constituent elements to nitrogen solute separates the homogeneous γ-phase crystalline structure into two-phase zone structure. Hence, the interface between two nano-crystalline structure distorts by itself to form an irregular zone boundary.

Let us summarize the microscopic view on the nitrided layer at the vicinity of the surface. Its microstructure consists of the γ-phase single-crystal structure and the γ_N/α_N, two-phase polycrystalline structure. The former structure is sparsely formed in the nitrided layer by plastic straining with nitrogen supersaturation. The fine grain boundaries with co-orientation of (111) are also formed along (111) slipping planes so that no nitrogen solute is present in its inside and it diffuses to the depth through these fine and larger grain boundaries. The latter structure is common to the nitrogen supersaturated zones. Each zone is composed of nitrogen-rich, γ-phase crystal and nitrogen-poor, α-phase one. This formation of the two-phase nanostructure is induced by different chemical compatibility of nitrogen solute to Cr and {Fe, Ni} in local. The local disturbance of nitrogen solute content triggers the local phase separation; this continuously propagates into the depth of the nitrided layer to form the fine two-phase structure.

Remember that the nitrogen supersaturated zones are advancing into the A- and B-grains in **Figure 9**. The highly strained zones are present in neighboring to the fine, two-phase zones. This local area in **Figure 9** is just resembling the microstructure at the surface of nitrided AISI316 in **Figure 15**. The γ-phase single-crystals are yielded by highly plastic straining in the similar manner to the fine distribution of high KAM zones in **Figure 9b**. Each refined g-phase single crystal in **Figures 9** and **15** is formed by enclosure of slipping lines with the orientation (111). The fine two-structured poly-crystals are also formed just near these high KAM zones in **Figure 9c**. Under high nitrogen enrichment through the single crystal zones, a two-phase nanostructure is formed by nitrogen supersaturation with its different chemical compatibility.

STEM analysis was utilized to describe the microstructure below NFE. In the mesoscopic analysis by EBSD, the α-phase zones, the modified grains, and the plastic strains are observed even below NFE in **Figure 10**. STEM analysis provides proof of microstructure modification by plastic straining. **Figure 18** shows the HAADF, ABF, and LAADF images below NFE. In the low-resolution imaging, the original AISI316 grain is divided into several subgrains in correspondence to the IPF profiles in **Figure 10c**. To be noticed in the HAADF image, this subgrain boundary has almost (111) plane; these subgrains are formed by the plastic straining in the easiest slipping lines. In addition, the inside of the subgrain also has slip-lines in (111) direction as seen in both ABF and LAADF images. This cross-slipping in (111) direction is just corresponding to the formation of γ-phase single-crystals in **Figure 16**. When the plastic straining co-works with the nitrogen supersaturation

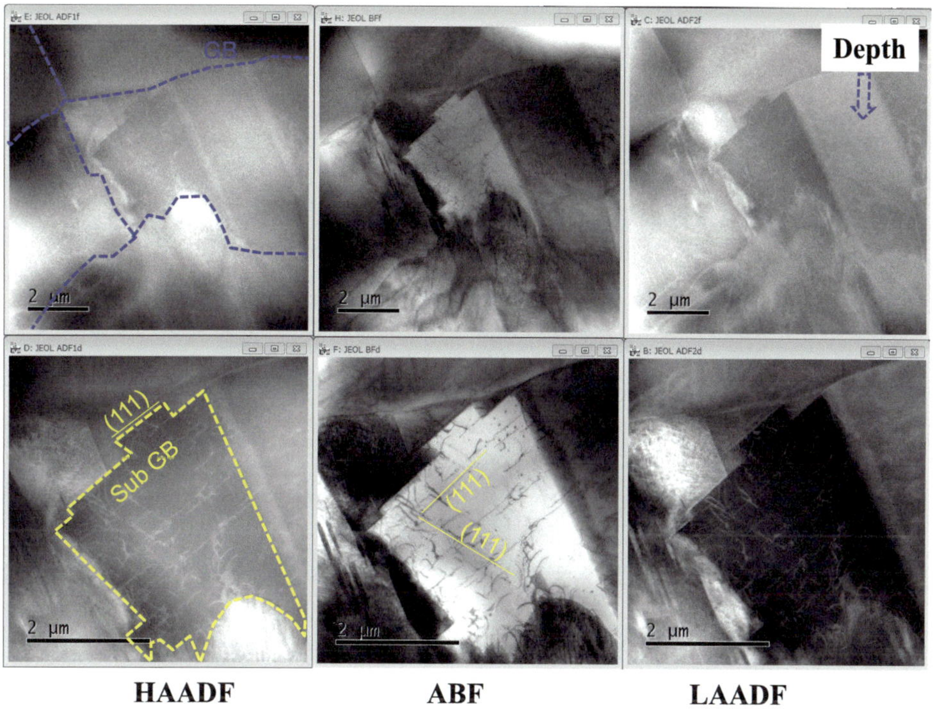

Figure 18.
HAADF, ABF, and LAADF analysis on the microstructure of the nitrided AISI316 just below NFE.

and boundary-diffusion processes, the microstructure in **Figure 18** is further modified and refined by the cross-slipping in plastic straining to be a single-crystal with the zone boundaries in (111) directions. Since the nitrogen content is nearly zero below NFE, the synergic relationship among the nitrogen supersaturation, the boundary diffusion, the plastic training, and the microstructure refinement stops intermediately to leave the plastically strained granular structure below NFE.

6. Application of nitrided AISI316 to surgery wire

A surgery wire requires high strength enough to apply the higher force for linkage of bones and tissues and for handling the miniature knife and tweezers. In addition, a surgery wire is demanded to have the high surface hardening enough to be free from damages and defects in mechanical troubles. Since the alloying elements for metallic medical parts and tools are strictly regulated, very few methods are useful to improve their mechanical properties and performance in practice. LT-PN is one of the most suitable methods to make surface treatment of AISI316 surgery wires without change of their constituent elements and element concentrations. In addition, the nitrogen supersaturated layer or NHSS layer has chemical inertness and sufficient corrosion toughness. No deterioration is expected to occur even by chemical polishing and pasteurization.

Figure 19 depicts the experimental procedure from the preparation of fine-grained AISI316 (FG-AISI317) wire to its uniaxial tensile testing. The FG-AISI316 wires with the average grain size of 1.5 µm, the diameter of 2.6 mm, and the length of 200 mm were prepared for LT-PN at 623 K for 14.4 ks. Other plasma processing conditions were the same as stated in Section 2. After nitriding, the side surface of wires is homogeneously nitrogen supersaturated to have a thick nitrided layer.

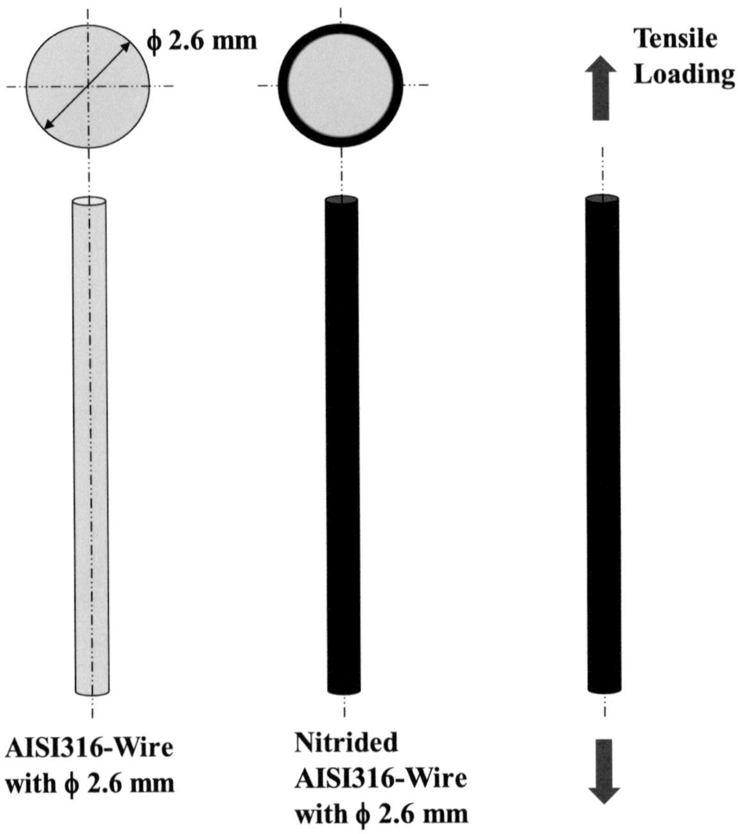

Figure 19.
An experimental procedure from the preparation of bare FG-AISI316 wire to the uniaxial tensile testing.

Mesoscopic analysis with the use of SEM–EDX is performed to verify the homogeneous nitriding behavior. The uniaxial tensile testing system (AUTOGRAPH AGS-X 10 kN; Shimazu Co., Ltd., Tokyo, Japan).) is employed to measure the applied stress to stroke relationship both for bare and nitrided FG-AISI316 wires. This uniaxial loading was terminated when the maximum applied load reached 6 kN before fatal ductile fracture for microstructure analysis.

The FG-AISI316 wire was uniformly nitrided to have the nitrogen supersaturated layer with the thickness of 40 μm; the un-nitrided FGSS316 matrix is continuously capped by this nitrided surface layer. Since this nitrided layer thickness is nearly equal to 40 μm in the nitrided FGSS316 plates at 623 K in **Figure 14**, the nitrogen supersaturation process advances from the circumferential surface of wire to the depth in a similar manner to low-temperature nitriding in the FG-AISI316 plates.

Figure 20 depicts the nitrogen solute mapping and IFP mapping in the circumferential and longitudinal cross-sections of nitrided FG-AISI316 wire, after uniaxial tensile loading. With respect to the nitriding mapping in both cross-sections, the nitrogen solute distribution is nearly the same as as-nitrided FG-AISI316 as compared between **Figures 12, 20a** and **c**. **Figure 20b** and **d** prove that the super-fine grained crystalline state with the two-phase structure in the nitrided FG-AISI316 is sustained during the uniaxial loading. To be noticed, the original equiaxed AISI316-matrix grains are elongated to be fibrous in the longitudinal direction as shown in **Figure 20d**. In addition, the FG-AISI316 matrix in the circumferential cross-section has bundle structure where each bundle of fibrous grains has specific crystallographic orientation as depicted in **Figure 20b**.

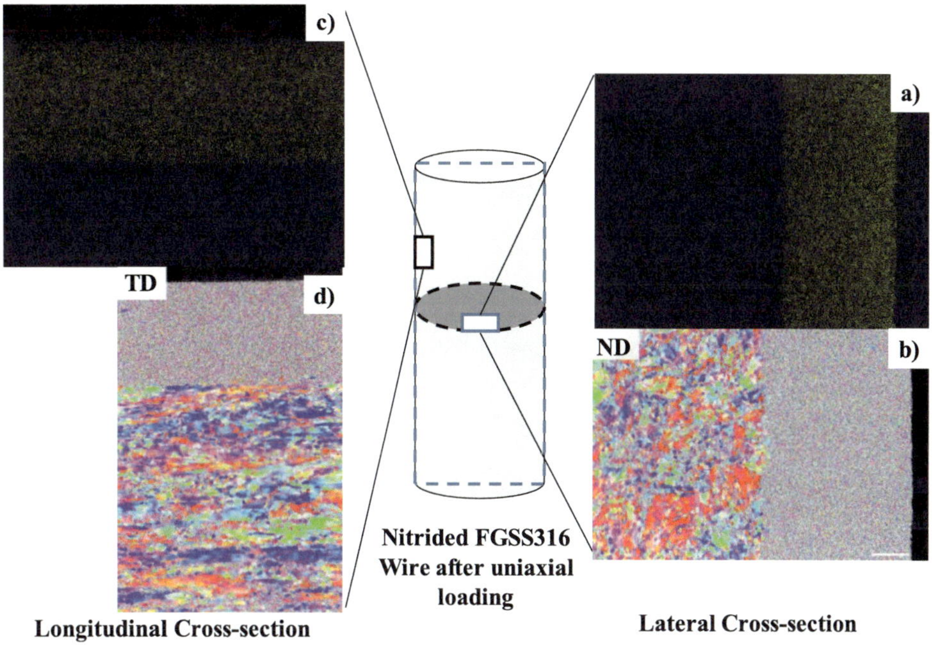

Figure 20.
Microstructure of the nitrided FG-AISI316 after uniaxial tensile testing. a) Nitrogen mapping in the lateral cross-section, b) its IPF profile, c) nitrogen mapping in the longitudinal cross-section, and d) its IPF profile.

This change of crystallographic structure in the FG-AISI316 matrix after uniaxial tensile loading, reveals that each original FG-AISI316 grain is elongated to fibrous grain by tensile loading and its crystallographic orientation is gradually rotated and aligned to the loading directions under the mechanical constraint by the nitrided layer. In particular, these fibrous grains are gradually sheared under this constraint to form a bundle structure with nearly the same crystallographic orientation. Let us investigate the effect of this crystallographic change in matrix on the mechanical properties of nitrided wire before and after uniaxial tensile loading.

Figure 21 compares the hardness profile on the lateral cross-section of nitrided wire before and after loading. The nitrided layer before uniaxial loading has a hardness of 1400 HV in agreement with the average hardness of the lateral cross-section of the nitrided FGSS316 plate in **Figure 13a**. This hardness abruptly decreases from 1400 HV to 400 HV at the nitriding front end of 40 μm; this hardness of 400 HV becomes constant toward the center of the inner matrix. After uniaxial loading, the matrix hardness remains the same as 400 V; the work hardening is not enhanced by this uniaxial tensile loading to accumulate the inner strains. On the other hand, the high hardness in the nitrided layer is further enhanced to be 1600 HV. This increase of hardness in the nitrided layer by uniaxial loading only corresponds to the further microstructure evolution and phase transformation from γ_N-phase to α_N-phase in the nitrided layer. That is, the microstructure change is locally induced in the nitrided layer by the applied plastic straining during the uniaxial loading.

The uniaxially applied stress (σ_{app}) to stroke (δ) relationship is compared between the bare and nitrided FG-AISI316 wires in **Figure 22**. The stiffness (K) is defined as the average change of the applied stress to the measured stroke up to $\delta = 1$ mm; e.g., $K = \sigma_{app}/\delta$. In the original FG-AISI316 wire, its stiffness becomes $K_0 = 500$ MPa/mm before nitriding; while the nitrided FG-AISI316 wire has a slightly greater stiffness (K_N) than K_0; e.g., $K_N = 580$ MPa/mm. The present nitrided wire is presumed as a composite of the nitrided layer with stiffness K_1 and

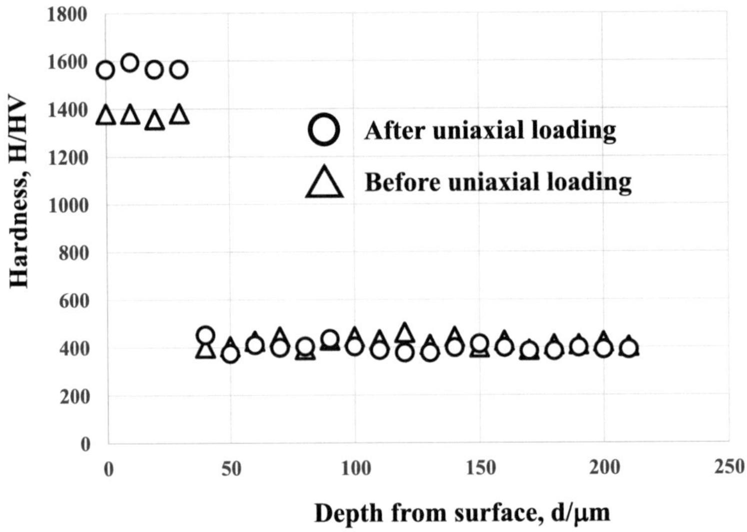

Figure 21.
Comparison of the hardness profile on the lateral cross-section of nitrided FG-AISI316 wires before and after the uniaxial tensile testing.

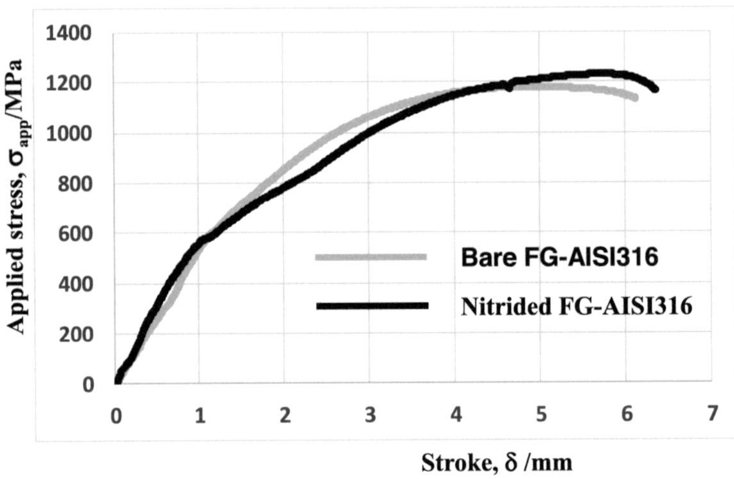

Figure 22.
Comparison of the applied stress to stroke relationship between the bare and nitrided FG-AISI316 wires in the uniaxial tensile testing.

the FG-AISI316 matrix with K_0. After the rule of thumb for the stiffness of composite materials [29], K_N is simply estimated by $K_N = (1 - f) \times K_0 + f \times K_1$, where f is the area fraction of the nitrided layer on the cross-section of the wire. In the present case, this f is only 6% because of the nitrided layer thickness of 40 μm of the wire lateral cross-section with a diameter of 2.6 mm. Assuming that $K_1 = (1600\ HV/400\ HV) \times K_0$, $K_N = 1.19 \times K_0 \sim 590$ MPa. This implies that the FG-AISI316 matrix in the nitrided wire is elastically constrained by the nitrided surface layer with higher hardness and stiffness similar to the lateral composite material.

7. Discussion

In HT-PN, the nitride layer of AISI316 was characterized by fine precipitation of ion and chromium nitride into the AISI316 matrix without its microstructure modification [7–10]. In bulk HNSS, the AISI316 matrix was supersaturated by nitrogen solute with is low content to stabilize the g-phase structure [6]. LT-PN of AISI316 substrate is characterized by its multi-dimensional structure. In the macroscopic view, the homogeneously nitrided AISI316 plates and wires at 673 K and 623 K have a nitrogen supersaturated layer or HNSS layer with an average nitrogen content of 4 mass% and an average hardness of 1400 HV. The initial γ-phase structure changes to be γ_N-/α_N, two-phase structure. Through the mesoscopic view, the nitrogen supersaturation process with two-phase structuring in the macroscopic view is further understood to investigate the synergic role among the nitrogen superstation, the plastic straining, the two-phase formation, the microstructure refinement, and the nitrogen boundary-diffusion. EBSD analysis on the heterogeneously nitrided matrix below NFE reveals that this synergic process advances to the localized grains near the nitrogen diffusing boundaries. This heterogeneous process is changed to be homogeneous by increasing the holding temperature and refining the initial grain size. In particular, the nitrided layer with the plateau of high nitrogen content by 4 mass%, the high hardness of 1400 HV, the fine two-phase microstructure, and the high plastic strains advances into the depth by the assistance of nitrogen boundary diffusion network in LT-PN of AISI316 with refined grain size. Microscopic analysis reveals that the nitrided layer consists of the g-phase single crystals with the grain size of 5 nm and the g-phase Cr (N) – a-phase Fe/Ni (N) poly-crystals. This single crystal is yielded by plastic straining to have fine (111)-oriented grain boundaries in parallel to the easiest slipping directions. High KAM zones or highly plastic-strained zones in EBSD analysis correspond to these single crystals. Nitrogen solute and dislocations induced by plastic straining are not housed in these single crystals but they transport through these crystals to the depth of the matrix. Fine two-phase structure in IPF profile by EBSD analysis also corresponds to this fine poly-crystal of Cr (N) and Fe/Ni (N) zones. This fine phase separation by the chemical compatibility of nitrogen solute among Cr, Fe, and Ni proves that nitrogen supersaturation into AISI316 induces the phase separation from g-phase AISI316 to two distinct phases with a locally neighboring system of nitrogen-rich and nitrogen-poor zones. A very fee study [30] in the literature reports that phase-separation selectivity is enhanced by nitrogen doping. In the present nitrogen-induced phase separation, high nitrogen solute content with its average of 4 mass% disturbs locally with separation to nitrogen-rich and –poor phases. Remembering the IPF profile in **Figures 8, 10, 14**, and **20**, where this two-phase structure forms a thick layer from the surface to the depth of nitrided AISI316. This local phase separation and nitrogen disturbance occur in a large volume with the advancement of nitrogen supersaturation and diffusion into the depth. The fine network of nitrogen zone- and grain-boundary diffusion paths in **Figures 16** and **17** sustains this large-scaled phase separation and nitrogen solute disturbance.

Localization of the nitrogen supersaturation and diffusion below NFE in **Figures 8** and **9**, and plastic straining effect on the formation of subgrains below NFE in **Figure 18**, also teaches the important role of synergic relation of plastic straining to nitrogen supersaturation and diffusion.

The mode-change from heterogeneous to homogeneous nitriding processes is controlled by refining the initial grain size of AISI316 substrates. In HT-NT, the nitriding behavior is sensitive to the holding temperature and chromium content; it has nothing to do with the microstructure. No studies were reported in the literature on LT-PN; This initial grain size effect on the inner nitriding is essential to understand the nitriding mechanism at low temperature. In the previous studies on HT-PN and LT-PN, the inner nitriding process is mainly dependent on the nitrogen body-diffusion process [7, 9]; hence, both HT-PN and LT-PN were thought to have nothing to do with the grain size in the microstructure. The mode-change by refining the initial grain size implies that the nitrogen solute at the lower holding temperature does not diffuse through the lattices in microstructure but diffuse through the zone and subgrain boundary network.

Assuming that each grain geometry with its size of D is modeled by a sphere with the diameter of D, its surface area is represented by πD^2. Let us calculate the number (N) of grains in a unit cubic cell with the edge length of L; e. g., $N = (L/D)^3$. Then, the grain boundary area (A) in the cubic cell is estimated by $A = \pi D^2 N = \pi L^3/D$. When the initial grain size is refined from D_0 to D_1, the grain boundary extension rate (E_r) is calculated by (D_1/D_0). In the present experiment, D_0 = 15 μm and D_1 = 1.5 μm; E_r becomes 10, or the initial grain boundary extends to be 10 times larger than the original AISI316. This enlargement of the grain boundary diffusion network area is responsible for the initial grain effect on the mode change.

Once the homogeneous nitriding is triggered by this initial grain refining, this nitriding process is sustained by the synergic relation among the nitrogen supersaturation, the plastic straining, the two-phase structuring, the microstructure refining, and the nitrogen boundary diffusion. To be remembered in [15, 16, 22, 23], the textures in the intensely rolled stainless steels completely disappeared through LT-PN. The pre-existing refinement in the granular structure of substrates is a key point to make sustainable and homogeneous nitriding of stainless steels. Other factors, mechanically induced by pre-straining, have nothing to do with nitriding behavior. This suggests that pre-forging provides yield the refined surface specimen for this homogeneous and sustainable LT-PN.

The metallic parts and tools in the medical application require various engineering items to improve their performance in practical use. The nitrided austenitic stainless steels have superior chemical inertness and corrosion toughness to the bare AISI316 products in addition to their high hardness. As reported in [31, 32], the nitrided nickel-free and martensitic stainless steels have sufficient corrosion toughness even in NaCl and etching solutions. Furthermore, the nitrided layer of AISI316 has enough machinability to be finished by using the PCD (Poly-Crystalline Diamond) - chipped and CBN (cubic boron nitride) – chipped milling tools [33–35]. The NHSS layer of nitrided AISI316 products can be precisely finished to have functional surfaces and interfaces owing to this feature of nitriding.

AISI316 has many derivatives such as AISI316L and AISI316LN. The former is selected and used as a high-strength sheet for piercing and metal-forming to fuel injection orifices [36]. The latter is also utilized as a structural component in the design of experimental fusion reactors [37]. Instead of those traditional alloying designs on AISI316L, the nitrogen can be utilized to reduce the effect of carbon interstitials to the mechanical performance by the nitrided AISI316 layer. AISI316LN can be exchanged by the nitrided AISI316 to prevent the structural component surface from the severe damages by spontaneous emission of particles. NHSS-layered AISI316 works as a structural member to be working instead of AISI316L and AISI316LN. Further studies are necessary to describe the interstitial solute as an alloying element in the redesign of AISI316 derivatives in its family.

8. Conclusion

The high-density plasma nitriding process is redesigned by using the plasma diagnosis to be working under the optimum conditions. In particular, OES (Optical Emissive-light Spectroscopy) is employed to search for the most suitable nitrogen–hydrogen gas flow rate to attain the higher yield of N_2^+ ions and NH radicals for low-temperature nitriding. AISI316 specimen is nitrided at 673 K and 623 K under the optimized conditions to describe the nitrogen supersaturation process by using multidimensional analysis.

Macroscopic evaluation on the formation of the nitrided layer is performed by XRD and SEM–EDX. The nitrogen supersaturation process with its higher content than 4 mass% induces the γ-lattice expansion and forms a plateau in the nitrogen content depth profile. This lattice expansion only occurs at the nitrogen diffusing zones so that the strains become incompatible between the nitrogen supersaturated and unsaturated zones. The unsaturated zones are plastically strained to distort their constituent grains.

EBSD is employed to make a mesoscopic evaluation on holding temperature and initial grain-size effects on the nitriding behavior in the nitrided layer and below the NFE. The plastic straining, the microstructure refining, and the two-phase structuring advance in synergy with nitrogen supersaturation and diffusion processes through the nitrided layer. This synergetic relation among these processes is observed in the heterogeneous nitriding process at 623 K. The microstructure refinement and two-phase structuring co-work, at the vicinity of localized nitrogen diffusion paths, with the nitrogen supersaturation and diffusion processes. This synergic working among four processes is the same as that in homogeneous nitriding. Through this localized nitrogen diffusion and supersaturation processes, the matrix microstructure far below NFE is also modified heterogeneously.

This heterogeneous nitriding turns to be homogeneous with increasing the holding temperature from 623 K to 673 K since the nitrogen diffusion rate is enhanced to reduce the localization behavior. To be noticed, the synergic process in the above is common to those heterogeneous and homogeneous nitriding processes. By refining the initial grain size of AISI316 before nitriding, this heterogeneous nitriding at 623 K also turns to be homogeneous even at 623 K. In addition, no localization in nitrogen supersaturation and diffusion occurs below NFE. The microstructure below NFE becomes the same as the original matrix of the original grain refined AISI316. This implies that the localization in nitrogen supersaturation and diffusion processes is suppressed by the enlargement of nitrogen diffusion paths.

Microscopic analysis with the use of STEM is employed to describe the microstructure refining and two-phase structuring in the nitrided AISI316 at 673 K. A single-crystal and poly-crystal like zones are formed at the vicinity of its surface. STEM analysis on the former reveals that the g-phase grain is much refined to have the size of 5 nm and the orientation of (111) by the synergic process in nitriding. The original coarse grain is nitrogen-supersaturated with its high nitrogen content and completely simple-sheared in (111)-direction by plastic straining to form very fine single-grain with no dislocations and nitrogen solutes left in its inside. STEM analysis discloses the two-phase structuring mechanism. After the phase mapping in EBSD analysis, this two-phase structure is only defined as a fine mixture of α- and γ-grains. The nitrogen-rich zone with more compatibility to chromium forms one phase while the nitrogen-poor zone with less compatibility to iron and nickel becomes another phase. In addition, these two-phase poly-crystals formed in nearest neighboring to the γ-phase single crystals. Furthermore, the STEM analysis on the microstructure below NFE proves that the sheared polycrystalline grains by the plastic straining are formed even in the absence of nitrogen solute.

The above multidimensional analysis demonstrates that low-temperature plasma nitriding is driven by the synergic relation of plastic straining, microstructure refining, and two-phase structuring with the nitrogen supersaturation and zone-boundary diffusion processes. The homogeneous inner nitriding of initially fine-grained AISI 316 is self-sustained by this synergic effect not to ignite the localization in nitrogen supersaturation even at 623 K. This self-sustainable nitriding is attractive to make surface treatment of stainless steel medical parts. Especially, the nitrided AISI316 wire is straightforwardly utilized as a surgery wire by its uniform surface hardness and high loading capacity.

Instead of the conventional alloying design such as AISI316L and AISI316LN, this self-sustainable nitriding provides a new way of high nitrogen structured AISI316 to industrial and medical applications.

Acknowledgements

The authors would like to express their gratitude to Dr. M. Saito, Prof. H. Yoshida, Prof. Ikuhara (University of Tokyo), and Mr. H. Morita (Nano-Film Coat, llc.) for their help in experiments.

Conflict of interest

The authors declare no conflict of interest.

Author details

Tatsuhiko Aizawa[1*], Tomomi Shiratori[2], Tomoaki Yoshino[3], Yohei Suzuki[3] and Takafumi Komatsu[3]

1 Surface Engineering Design Laboratory, SIT, Tokyo, Japan

2 University of Toyama, Toyama, Japan

3 Komatsu-Seiki Kosakusho, Co., Ltd., Suwa, Japan

*Address all correspondence to: takafumi@komatsuseiki.co.jp

References

[1] Cobb HM. The History of Stainless Steel. Materials Park, OH: ASM International; 2010

[2] Peckner D, Bernstein IM. Handbook of Stainless Steels. New York: McGraw Hill; 1977

[3] Stainless Steels for Design Engineers. ASM International. 2008

[4] Available from: https://www.oshwin.com/blog/difference-between-316-and-316l.html [Retrieved at 2021/12/1]

[5] Wu C, Li S, Zheng C, Wang X. Microstructural evolution in 316LN austenitic stainless steel during solidification process under different cooling rates. Journal of Materials Science. 2016;**51**:2529-2539

[6] Imai Y, Murata T, Sakamoto M. High Nitrogen Steels. Agne-Publisher; 2005

[7] Kuwahara H. Surface modification of iron alloys by plasma nitriding and carburizing [PhD Thesis]. Kyoto University; 1992

[8] Anzai M, editor. Heat and Surface Treatment of Die Materials. Tokyo, Japan: Nikkan-Kogyo-Shinbunsha; 2011

[9] Hiraoka Y, Inoue K. Prediction of nitrogen distribution in steels after plasma nitriding. Denki-Seiko. 2010;**86**: 15-24

[10] Aizawa T, Sugita T. High density RF-DC plasma nitriding of steels for die and mold technologies. Research Reports SIT. 2013;**57**(1):1-10

[11] Bell T. Surface engineering of austenitic stainless steels. Surface Engineering. 2002;**18**:415-422

[12] Borgioli F, Galvanetto E, Bacco T. Low temperature nitriding of AISI300 and 200 series austenitic stainless steels. Vacuum. 2016;**12**:51-60

[13] Aizawa T. Low temperature plasma nitriding of austenitic stainless steels. Chap. 3. In: Stainless Steels. London, UK: InTechOpen; 2019. pp. 31-50

[14] Aizawa T, Yoshihara S-I. Homogeneous and heterogeneous micro-structuring of austenitic stainless steels by low temperature plasma nitriding. IOP Conference Series: Materials Science and Engineering. 2018;**372**:012059

[15] Aizawa T, Yoshino T, Shiratori T, Yoshihara S-I. Grain size effect on the nitrogen supersaturation process into AISI316 at 623 K. ISIJ International. 2019;**59**:1886-1892

[16] Aizawa T. Advanced technologies for die materials and manufacturing. Bulletin, JSTP. 2021;**4**(41):291-296

[17] Farghali A, Aizawa T, Yoshino T. Microstructure/mechanical characterization of plasma nitrided fine-grain austenitic stainless steels in low temperature. Nitrogen. 2021;**2**:244-258

[18] Benda M, Musil J. Plasma nitriding enhanced by hollow cathode discharge – A new method for formation of superhard nanocomposite coatings on steel surfaces. Vacuum. 1999;**55**(2): 171-175

[19] Yunata EE. Characterization and application of hollow cathode oxygen plasma [PhD Thesis]. SIT; 2016

[20] Aizawa T, Rsadi I, Yunata EE. High density RF-DC plasma nitriding under optimized conditions by plasma diagnosis. Journal of Applied Sciences. 2021 (in press)

[21] Domain C, Becquart CS, Foct J. Ab initio study of foreign interstitial atom

(C, N) interactions with intrinsic point defects in a-Fe. Physical Review B. 2004;**69**:144122

[22] Aizawa T, Yoshihara S-I. Inner nitriding behavior and mechanism in stainless steels at 753 K and 623 K. The SEATUC Journal of Science and Engineering (SJSE). 2019;**1**:13-20

[23] Aizawa T, Yoshihara S-I. Inner nitriding behavior and mechanism in stainless steels type AISI316 at 623 K. In: Proc. 13rd SEATUC Conference (March 13th, 2019; Hanoi, Vietnam). pp. 121-127

[24] Kamaya M, Wilkinson AJ, Titchmarsh JM. Quantification of plastic strains of stainless steel and nickel alloy by electron backscatter diffraction. Acta Materialia. 2006;**54**:539-548

[25] Aizawa T, Shiratori T, Komatsu T. Micro—/nano-structuring in stainless steels by metal forming and materials processing. Ch. 1. In: Electron Crystallography. London, UK: IntechOpen; 2019. pp. 1-23

[26] Tomimura K, Takaki S, Tokunaga Y. Reversion mechanism from deformation induced martensite to austenite in metastable austenitic stainless steels. ISIJ International. 1991; **31**(12):1431-1437

[27] Aizawa T, Shiratori T, Komatsu T. Integrated manufacturing of fine-grained stainless steels for industries and medicals. Ch. 1. In: Engineering Steels and High-Entropy Alloys. London UK: IntechOpen; 2019. pp. 3-26

[28] Korchuganova AV, Tyumentsevab AN, Zolnikova KP, Litovchenkoab IY, Kryzhevicha DS, Gutmanasc E, et al. Nucleation of dislocations and twins in fcc nanocrystals: Dynamic of structural transformations. Journal of Materials Science and Technology. 2019;**35**(1): 201-206

[29] Nemat-Nasser S, Hori M. Micromechanics: Overall Properties of Heterogeneous Materials. Amsterdam, Netherlands: North-Holland; 1992. p. 113

[30] Yuan N, Chen J, Zhou H, Ali MC, Guan M, Qiu H. Nitrogen-doping to enhance the separation selectivity of glucose-based carbon dots –modified silica stationary phase for hydrophilic interaction chromatography. Talanta. 2020;**218**:121140

[31] Borgioli F, Galvanetto E, Bacci T. Corrosion behavior of low temperature nitrided nickel-free, AISI200 and AISI300 series austenitic stainless steels in NaCl. Corrosion Science. 2018;**136**: 352-365

[32] Aizawa T, Yoshihara S-I. Microtexturing into AISI420 dies for fine piercing of micropatterns into metallic sheets. Journal of Service Theory and Practice. 2019;**60**(698): 53-57

[33] Aizawa T, Fukuda T. Microstructure and micro-machinability of plasma nitrided AISI420 martensitic stainless steels at 673 K. In: Top 5 Contributions in Materials Science. London: Avid Science. 2019;**2**(6):1-24

[34] Aizawa T, Morita H, Fukuda T. High machinability of plasma-nitrided HPM80 dies at 673 K by PCD-tools for hot mold-stamping. Procedia Manufacturing. 2020;**47**:725-731

[35] Aizawa T, Morita H, Fukuda T. Nitrogen supersaturation into AISI420 mold for precise machining. In: Proc. 23 ESAFORM Conf. 2022 (In press)

[36] Aizawa T, Sato T, Shiratori T. Micro-joining of shaped stainless steel sheets for fuel injection orifice with high misting capability. In: Proc. 22nd Int. ESAFORM Conf. AIP Conf. Proc. 2019. 2113 pp. 050007-1 – 050007-6

[37] Ferraria L, Palmieria A, Pepatob A, Prevedellb A, Dimab R, Udupb E. Production quality controls and geometric characterization of the IFMIF-RFQ modules via the usage of a coordinate measuring machine. Fusion Engineering and Design. 2017;**115**:23-32

Chapter 9

Study on the Perspective of Mechanical Properties and Corrosion Behaviour of Stainless Steel, Plain and TMT Rebars

Indrajit Dey, Pallabi Manna, Muralidhar Yadav, Nisith Kumar Tewary, Jayanta Kumar Saha and Swarup Kumar Ghosh

Abstract

In the present research, the effects of various alloying elements and microstructural constituents on the mechanical properties and corrosion behaviour have been studied for four different rebars. The microstructures of stainless steel and plain rebar primarily reveal equiaxed ferrite grains and ferrite-pearlite microstructures, respectively, with no evidence of transition zone, whereas tempered martensite at the outer rim, followed by a narrow bainitic transition zone with an internal core of ferrite-pearlite, has been observed for the thermomechanically treated (TMT) rebars. The hardness profiles obtained from this study display maximum hardness at the periphery, which decreases gradually towards the centre, thereby providing the classical U-shaped hardness profile for TMT rebars. The tensile test results confirm that stainless steel rebar exhibits the highest combination of strength (≈755 MPa) and ductility (≈27%). It has been witnessed that in Tafel plots, the corrosion rate increases for all the experimental rebars in 1% HCl solution, which is well expected because the acid solutions generally possess a higher corrosive environment than seawater (3.5% NaCl) due to their acidic nature and lower pH values. However, all the experimental results obtained from Tafel and Nyquist plots correlate well for both 1% HCl and 3.5% NaCl solutions.

Keywords: stainless steel, reinforced bar, TMT, microstructure, mechanical properties, corrosion behaviour

1. Introduction

For the past few years, an inspection of reinforced concrete structures with steel rebars has become a subject of research because concrete provides an alkaline environment that is suitable for the spontaneous passivation of steel rebars. These steel rebars act as a reinforcement when tensile loads are applied [1]. But, still, there is a possibility of corrosion in these steel rebars when aggressive chloride ions are present in the environment [1]. Corrosion in rebars may result in surrounding concrete cracking, reduction in the bond strength of the concrete and the rebars and lowering the combination of strength and ductility [1–3].

Extensive research has been carried out to date on corrosion in steel rebars but a modest solution can be the usage of stainless steels rebars as a partial or complete replacement of the reinforcements in concrete. This solution is economically viable because there may be a higher initial investment cost but this can be compensated by low-repairing cost, less maintenance effort and most importantly long-range services of the designed structures [4]. As per earlier studies, it is well established that austenitic stainless steels can be considered as the common choices for rebars, whereas there is a wide range of applications of the ferritic and the duplex stainless steels that include structural components in the construction industries [1, 5]. However, it cannot be domineered that the austenitic stainless steels are expensive due to the higher content of elements such as Ni and Mo. It has been reported earlier that stainless steel eliminates chances of corrosion but does not rely on concrete when it is subjected to the ingress of chlorides from marine environments. For the last few years, plenty of research has been carried out in search of more economically friendly novel stainless steels as reinforcements [6–12]. It is also true that, since the cost of these steels can be compared with the cost of common black steel reinforcement rather than the cost of highly alloyed stainless steels, the use of low alloyed stainless steel reinforcement has been accepted as economically more convenient [6].

Thermo-mechanical treatment (TMT) has gained researchers' attention for producing high-strength steels with lean chemistry at a reduced cost with excellent mechanical properties [13–16]. In TMT, simultaneous application of heat and deformation causes microstructural refinements [16–18]. These materials are highly used in the construction sectors for the construction of dams, bridges, buildings, flyovers and also in various other structural materials [16, 19]. The durability of these TMT rebars is an immense problem due to corrosion in reinforced concrete structures, which is needed to be minimized as much as possible to enhance the life of that structure. Substandard quality rebars can also cause damage to the towers, buildings and other constructions in the seismic zones. However, in reinforcement concrete, the presence of alkaline solution protects the TMT rebars from corrosion due to the formation of a passive layer over the steel surface [20–22]. However, encountering chloride solution in this structure causes the breakdown of this passive layer. In the coastal and marine areas, excessive level of chloride causes problems of chloride-induced corrosion, thereby creating problems in corrosion resistance of the TMT rebars in the concrete [23–27].

It is well established that corrosion of steel reinforcement bars primarily depends on the stability of different phases that are likely to form in the cement paste and the effectiveness of the oxide layer that forms on the steel bar surface, that is, the passive layer [28]. It has been reported earlier that the concrete pore solutions preserve a strongly alkaline medium due to the significant filling of calcium hydroxide in the concrete pores depending on the hydration reactions of cement. This alkaline environment is suitable for the formation of a stable passive film on the rebar surface and thereby provides significant protection to the steel rebar against corrosive environments [28]. However, chloride-induced corrosion has been reported as a primary source for the local breakdown of this passive film on the surface of the steel bars [28–30]. It is also known that the volume of rebar increases when the corroded product forms on the rebar surface and results in cracks in the concrete, thereby causing failure [27].

This paper primarily emphasizes the study of mechanical and corrosion behaviours of different types of rebars in two different aqueous solutions. It is expected that this study will provide valuable information regarding the improvement in the design parameters and also the life cycle cost calculation of various structures as well as towards the selection of the best-suited rebars for industrial and marine construction purposes.

2. Experimental procedure

Four different steel rebars such as Fe-600, galvanized steel, stainless steel and plain rebars were chosen for this study. The approximate diameters of these samples were 12 mm, 8 mm, 16 mm, and 11 mm, respectively. Notably, Fe-600 and galvanized rebars are thermomechanically treated, whereas the other two (stainless and plain) rebars were not subjected to thermomechanical treatment. The nominal chemical compositions (wt. %) of all the rebar samples are given in **Table 1**. First, the samples of suitable size were cut from these steel rebars as per the requirement of optical, hardness, tensile and corrosion tests. All the samples were then tested in an optical emission spectrometer (Thermo 3460) to analyse the chemical compositions of the rebars. Standard metallographic sample preparation techniques such as grinding (belt and paper) followed by polishing (coarse, fine and diamond) and etching by using 2% nital and Snyder solutions were used to observe the microstructural constituents under an optical microscope (Leica) and scanning electron microscope (SEM).

The etched samples were further used for the Vickers Micro-Hardness Testing (Leica-VMHT) under 300 gf load and 20 seconds dwell time to analyse micro-hardness profile throughout the cross sections with a particular interval of distance. Hardness was measured on the sample surface as a function of distance from the edge to edge, keeping an interval of 50 μm, and the recorded variation in hardness values was then plotted. It is important to mention here that the hardness profile instead of surface hardness was recorded because the hardness values were expected to decrease from the periphery to the centre. A universal testing machine (UTN-10) was utilized to conduct the tensile test with a crosshead speed of 4.5 mm/min and a gauge length of 65 mm.

The electrochemical measurements were performed using a working electrode (embedded rebar specimen); a counter electrode (graphite) was placed to one side of the rebar specimen and also an independent reference electrode; that is, saturated calomel electrode (SCE) was used to observe the corrosion behaviour. Precise electrode placement was not critical, since the conductivity of the electrolytes was high. The equipment used to observe the corrosion behaviour of the rebar samples was Origalys Potentiostat, combined with the Origamaster 5 module. Electrochemical impedance spectroscopy (EIS) was utilized in the frequency range of 100 kHz–100 MHz with an AC voltage of 5 mV. The characteristic capacitance of this frequency band was maintained in between 10^{-9} and 10^{-6} F/cm^2 [10]. The rebar samples of standard size (area) were prepared and then immersed in two different types of solution of 3.5% NaCl and 1% HCl. These particular concentrations were chosen to simulate the saline conditions possible in marine and acidic rain in industrial areas [31, 32]. In this context, it is important to mention that the area was ≈76 mm^2 for Fe 600 rebar, whereas the same for the other three samples was kept at ≈75 mm^2 for the EIS study.

3. Results and discussion

3.1 Alloy composition

Table 1 shows the compositional analysis of the different types of TMT rebars (IS 1786: 2008 for Fe 600 and IS 12594:1988 for galvanized). It has been already reported that an increase in yield strength of reinforcing bars occurs by raising carbon as well as manganese content or by microalloying. It is also true that higher carbon content may cause lower weldability and ductility [13, 16, 23]. It is worth mentioning here that all the steel rebar samples selected in the present investigation contain a lower amount of carbon since a higher amount of carbon can cause harmful carbide formation and pearlite formation in the as-rolled steels thereby causing micro-galvanic corrosion [23, 33]. Major alloying elements other than carbon are manganese and silicon in the case of all the rebar samples. However, in the case of stainless steel rebar (IS 16651:2017), apart from these two elements, chromium and nickel are also present to enhance the 'stainless' property of this steel by forming a stable and protective oxide layer to make them corrosion-resistant [34–36]. In this context, it is imperative to mention here that the corrosion resistance of the stainless steel rebar depends on the formation of the chromium layer ($\approx$5 nm thickness) and the content of chromium that keeps the rebar surface electrochemically passive in corrosive environments [37]. It is well known that manganese increases the hardenability and tensile strength of the steel [23, 33]. The addition of silicon is beneficial to increase the hardness and strength. It also acts as a deoxidizer during the casting of the rebar, thereby helps in reducing the casting defects and allows sound steel castings [23, 38].

3.2 Microstructure analysis

3.2.1 Optical micrographs

Figure 1 shows the optical micrographs of different rebar samples (cross section). **Figures 1(a)** and **(b)** reveals the optical micrographs of the Fe 600 and galvanized rebar samples, respectively, that exhibit intermediate zones with three distinctly separate regions for both the samples. On the other hand, microstructures consisting of equiaxed ferrite grains for stainless steel (**Figure 1(c)**) and ferrite-pearlite type for plain rebar (**Figure 1(d)**) have been observed throughout the samples because these two rebars were processed through the conventional cold rolling process and not subjected to thermomechanical treatments (TMT). The presence of dark peripheral ring (edge/rim) consists of tempered martensite with an intermediate narrow transition zone of bainite with a comparably grey core with ferrite-pearlite microstructure as shown in **Figures 1(a)** and **(b)** [23, 33, 39, 40]. It is imperative to mention here that in the TMT process, quenching of hot-rolled bars

Rebar specimen	C	Mn	Si	Cr	Ni	S	P
Fe 600	0.30	0.23	0.16	—	—	0.04	0.035
Galvanized	0.22	0.58	0.23	0.03	—	0.04	0.035
Stainless steel	0.022	0.65	0.36	12.98	0.09	0.008	0.021
Plain	0.126	0.526	0.175	0.007	0.018	0.046	0.078

Table 1.
Chemical composition (wt. %) of all the experimental rebar specimens.

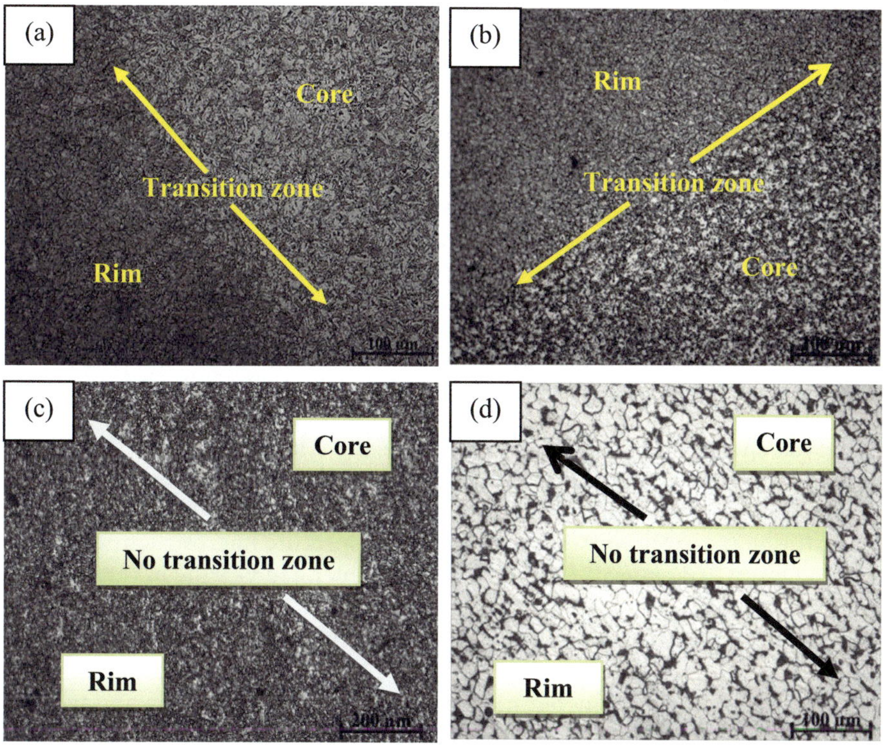

Figure 1.
Optical micrographs of different rebar samples (a) Fe 600, (b) galvanized rebar, (c) stainless steel and (d) plain rebar.

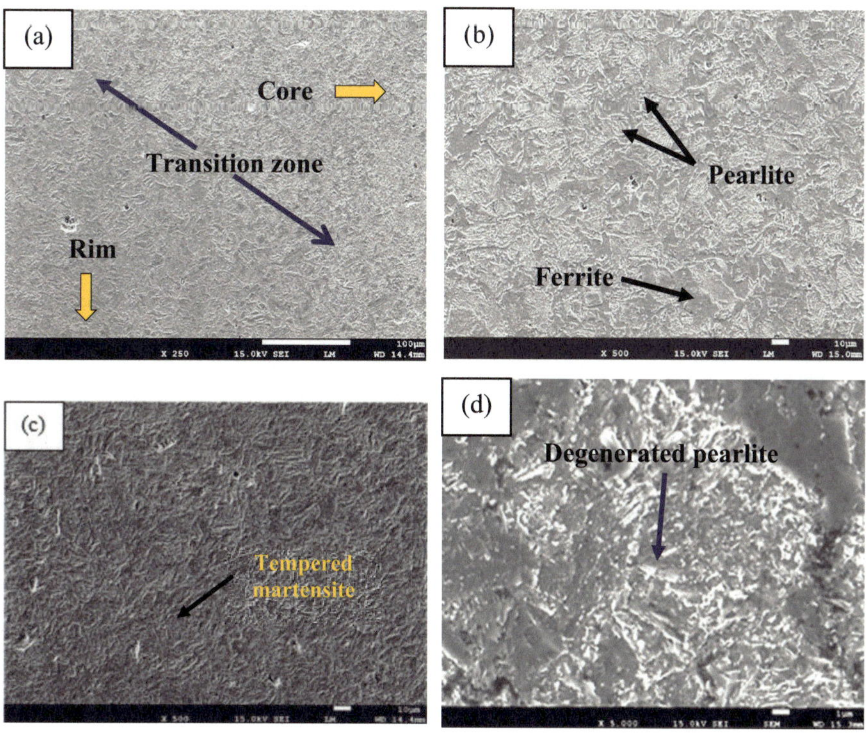

Figure 2.
SEM micrographs of Fe 600 rebar sample showing (a) transition zone, (b) core region, (c) outer rim region and (d) pearlite at higher magnification.

with water jet leads to the formation of martensite at the outer surface of the rebars. The residual heat flow across their bar section during cooling results in the tempering of the initially formed martensite with finer grains, which is self-tempering in nature and also produces ferrite-pearlite with coarser grains or mixed microstructure in the core region [13, 33, 41, 42]. A clear boundary separating the outer ring from the inner core is also visible in the case of TMT rebars (**Figures 1(a)** and **(b)**). This boundary can be considered as the demarcation line between the tempered martensite and ferrite-pearlite microstructures.

3.2.2 SEM micrographs

Figures 2 and **3** reveal the SEM microstructures of the Fe 600 rebar and galvanized rebar samples, respectively, consisting of the tempered martensitic rim at the periphery as shown in **Figures 2(c)** and **3(c)**, along with a transition zone consisting of bainite as shown in **Figures 2(a)** and **3(a)**, followed by a ferrite-pearlite mixed microstructure in the core as shown in **Figures 2(b)** and **3(b)**. In this context, it is noteworthy to mention here that the pearlite microstructure that is shown in **Figure 2(d)** reveals the presence of degenerated pearlite in the case of the Fe 600 rebar sample, which can be correlated with inadequate carbon diffusion during cooling. Detailed discussion on the formation of degenerated pearlite and bainite is available in the literature [23, 33, 39–44]. **Figure 3(d)** primarily reveals pearlite microstructure at a higher magnification with greyish ferrite and whitish cementite flake-like structures, which is completely distinguishable for galvanized rebar samples.

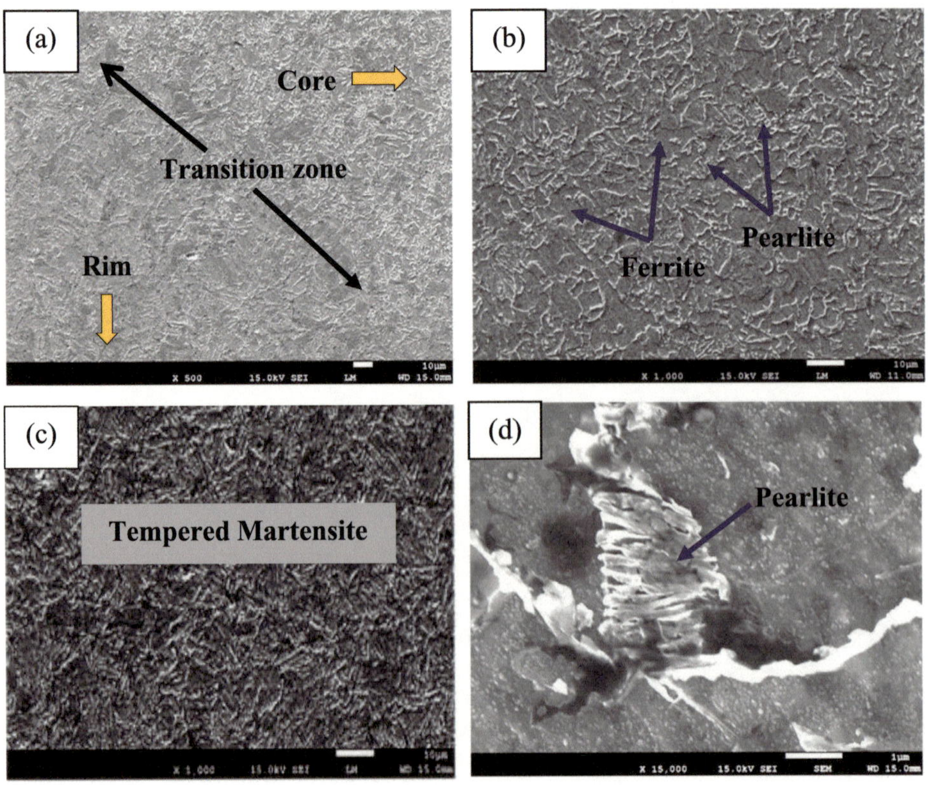

Figure 3.
SEM micrographs of galvanized rebar sample showing (a) transition zone, (b) core region, (c) outer rim region, and (d) pearlite at higher magnification.

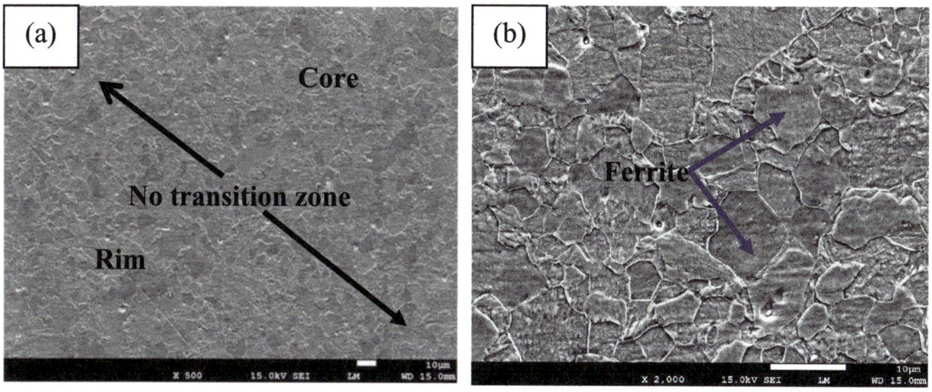

Figure 4.
SEM micrographs of stainless steel rebar sample showing (a) no transition zone and (b) higher magnification micrograph showing equiaxed ferrite grain.

SEM microstructures of stainless steel rebar and plain rebar samples are shown in **Figures 4** and **5**, respectively. These microstructures mainly consist of equiaxed grains of ferrite separated by clearly visible distinct grain boundaries in the case of stainless steel rebar sample as shown in **Figure 4(b)**. It cannot be overruled that as the carbon percentage is very low and chromium (ferrite stabilizer) percentage is high for the investigated stainless-steel rebar, hence the microstructure is ferritic. However, mixed ferrite-pearlite microstructures with no transition zone throughout the sample have been observed for plain rebar. The lamellar configuration of pearlite with some amount of resolved pearlite has been shown at a higher magnification at the inset of **Figure 5(b)**.

3.3 Mechanical properties

3.3.1 Hardness profile analysis

Figure 6 shows the hardness profiles of different types of steel rebar samples (along the cross-sectional diameter). It has been reported earlier that the hardness value of TMT rebar samples becomes maximum at the periphery due to the presence of tempered martensite, which gradually decreases towards the

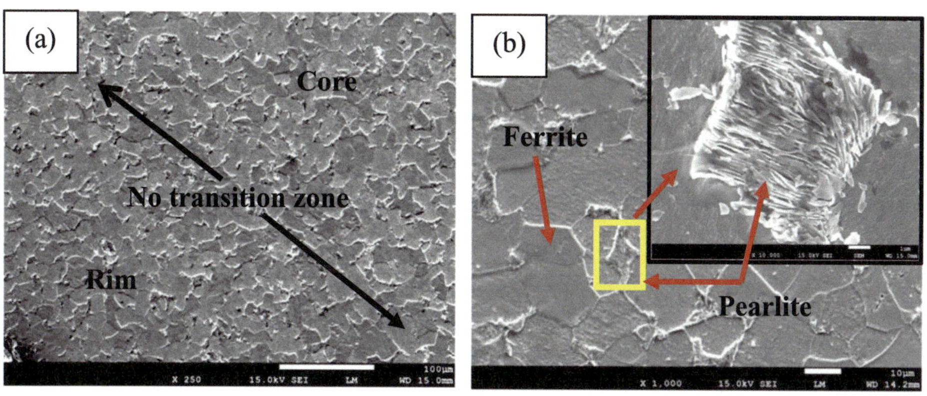

Figure 5.
SEM micrographs of plain rebar sample showing (a) no transition zone and (b) pearlite at higher magnification.

centre due to the formation of ferrite-pearlite mixed microstructure thereby exhibiting the classical U-shaped profiles [13, 23, 45]. It has been observed that the hardness value of Fe 600 and galvanized rebar samples in the periphery or outer rim region is around ≈285 HV, which is an indication of the presence of the tempered martensite or bainite phase [23, 45]. It has been also observed that the hardness value of the transition zone is ≈220 HV for both TMT rebar samples due to the presence of lower bainite, whereas for the core region or ferrite-pearlite region, the hardness value is ≈200 HV for Fe 600 rebar and ≈180 HV for galvanized rebar, respectively.

Figures 7(**a**) and (**b**) displays the comparative hardness variations for stainless steel and plain rebar samples, respectively. It is evident from **Figure** 7 that the stainless steel rebar shows higher values of hardness (≈195 HV to 260 HV) than the plain rebar (≈145 HV to 205 HV). Previous studies have confirmed that the hardness value in steel samples predominantly depends on the carbon equivalent value (CEV) [23]. It has been also reported that the weldability primarily depends on CEV for carbon steel; that is, higher CEV leads to a hard and brittle heat-affected zone (HAZ). The microstructure in the HAZ zone has an important role in the mechanical properties of the weldment; therefore, CEV is an important parameter for rebars [46]. **Table 2** summarizes the estimated carbon equivalent values of different rebar samples that were calculated by using Eq. (1). It can be seen in **Table 2** that the plain rebar sample shows the lowest CEV (≈0.2163), thereby showing the lowest value of hardness among all the samples. It is imperative to mention here that although the stainless steel rebar consists of only equiaxed ferrite microstructure (**Figure 4**), due to high carbon equivalent value and the presence of other alloying elements such as Mn, Si, Ni (**Table 1**), and it shows a higher hardness value. It cannot be domineered that Fe 600 steel rebar has a higher value of CEV (≈0.3383) than that of galvanized rebar (≈0.3227), which results in a higher core hardness value (≈196 HV) (**Figure 6**) when compared with that of galvanized rebar sample (≈184 HV). The aforesaid two TMT rebars with CEV < 0.42 according to IS 1786:2008 standard will exhibit superior weldability. However, the

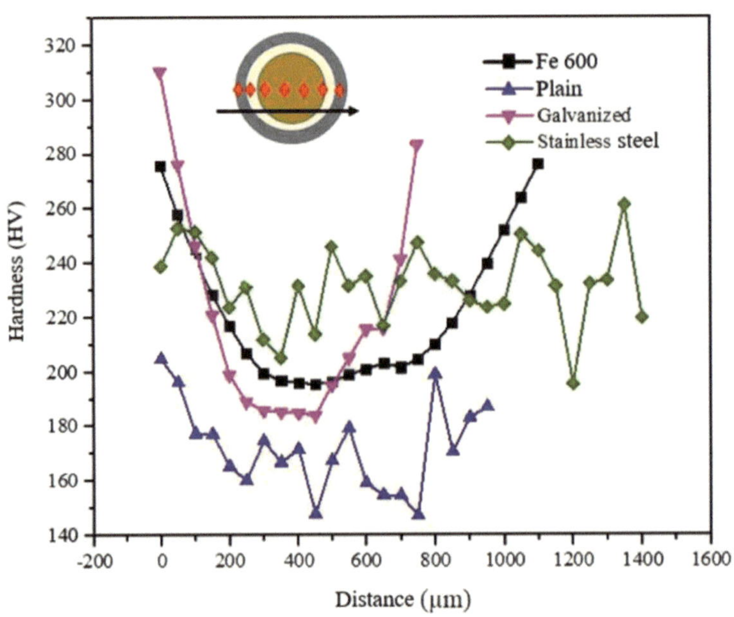

Figure 6.
A comparison of hardness data among all the rebar specimens.

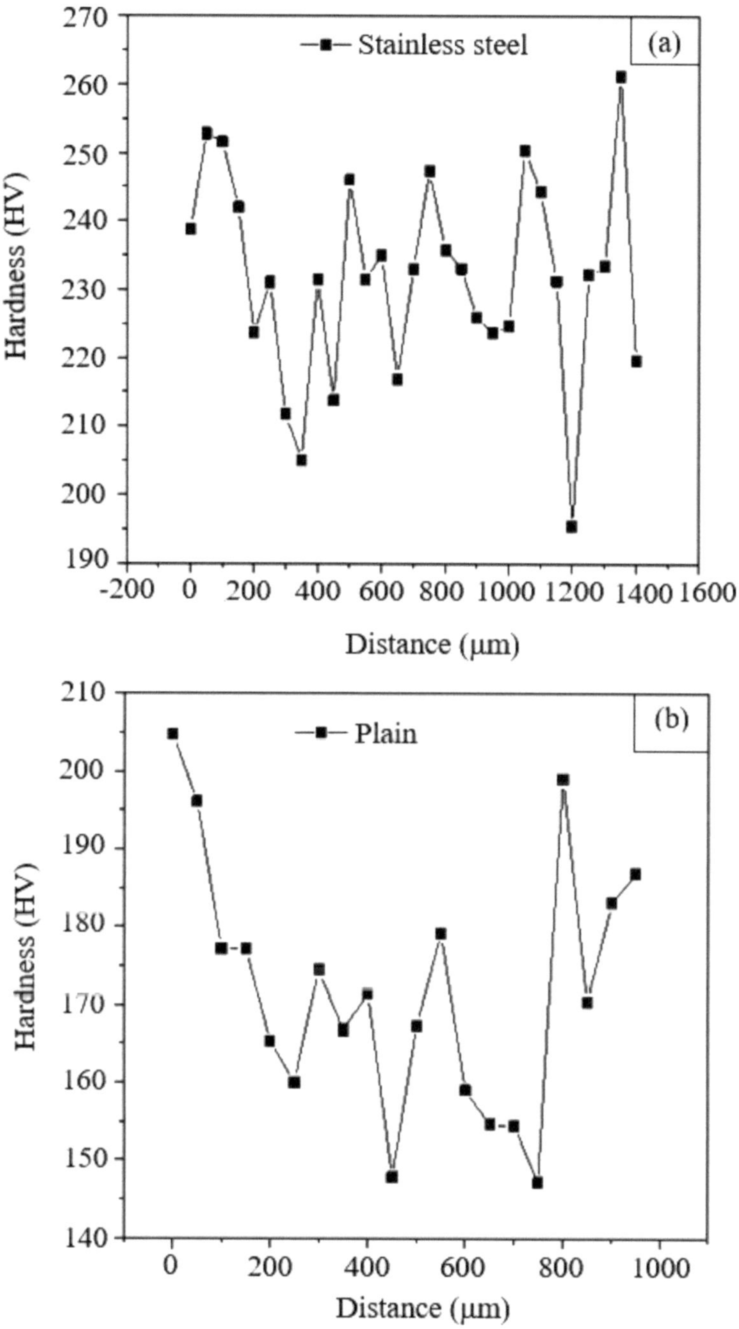

Figure 7.
Plots showing the variation of hardness with distance for (a) stainless steel rebar and (b) plain rebar samples.

Rebar specimen	Carbon Equivalent Value (CEV)
Fe 600	0.3383
Galvanized	0.3227
Stainless steel	2.7323
Plain	0.2163

Table 2.
Estimated carbon equivalent value (CEV) for different rebar samples.

stainless steel rebar (ferritic grade) shows higher CEV (≈2.7323) because of higher Cr content but it can be welded in accordance with IS 16651:2017.

$$CEV = C + Mn/6 + \{(Cr + Mo + V)/5\} + \left\{\frac{Cu + Ni}{15}\right\} \tag{1}$$

(Where all the elemental values are expressed in wt.%)

3.3.2 Tensile properties

Table 3 summarizes the values of yield strength (YS), ultimate tensile strength (UTS), UTS/YS ratio and ductility in terms of elongation percentage (% EL) of different rebars. From **Table 3**, it is evident that the highest combination of strength and ductility has been achieved for stainless steel, followed by Fe 600 rebar among all the experimental rebars. It is a well-established fact that a higher CEV leads to higher UTS variations (**Table 2** *vis à vis* **Table 3**) [23]. **Table 4** shows the estimated ring diameter and area of tempered martensite in TMT rebars. It is noteworthy to mention that a greater amount of tempered martensitic ring area leads to higher strength (**Table 3** *vis à vis* **Table 4**) in the case of Fe 600 rebar. The estimated ring area of tempered martensite (≈39.56%) in Fe 600 rebar results in higher strength and lower ductility compared with the galvanized rebars with ≈20.10% ring area. It has been also noticed from **Table 4** that the area percentage of tempered martensite ring increases with the increasing diameter of the tempered martensitic rim. Plain rebar shows the maximum ductility due to the lower carbon content, whereas stainless steel with the lowest carbon and higher Cr contents having a completely ferritic microstructure and ferrite being a soft phase exhibits somewhat lower ductility than plain rebar. From earlier research, it has been observed that in seismic zones according to the ASTM A706:2006, UTS/YS ratio should be kept ≈1.25 [23, 33]. Previous studies have confirmed that rebars with UTS/YS ratio ≈1.19–1.24 provide greater advantages to their fatigue life [23, 47]. Therefore, it is clear from **Table 3** that stainless steel rebar with UTS/YS value ≈1.20 can become a potential candidate for the seismic hazard zone, whereas the other experimental rebars with lower values of UTS/YS ratio are expected to show better load resistance capacity in the seismic zone [23, 48].

Rebar specimen	YS (MPa)	UTS (MPa)	(UTS/YS)	Elongation (%)
Fe 600	600	660	1.10	10
Galvanized	500	545	1.09	12
Stainless steel	630	755	1.20	27
Plain	409	470	1.15	29

Table 3.
Tensile properties of various experimental rebar specimens.

Rebar specimen	Tempered martensite ring diameter (mm)	Tempered martensite ring area (%)
Fe 600	1.38	39.56
Galvanized	0.67	20.10

Table 4.
Estimated ring diameter and area of tempered martensite in TMT rebars.

3.4 Corrosion behaviour

3.4.1 Polarization curves

Figures 8 and **9** display the Tafel plots of different rebars in acid (1% HCl) and seawater (3.5% NaCl) solutions. It can be seen from **Figures 8(a)** and **(b)** that in 1% HCl solution, stainless steel rebar undergoes passivation, while other rebars show a constant enhancement of anodic current density with increasing applied potential. Generally, the state of corrosion in rebars can be classified as passive, active or

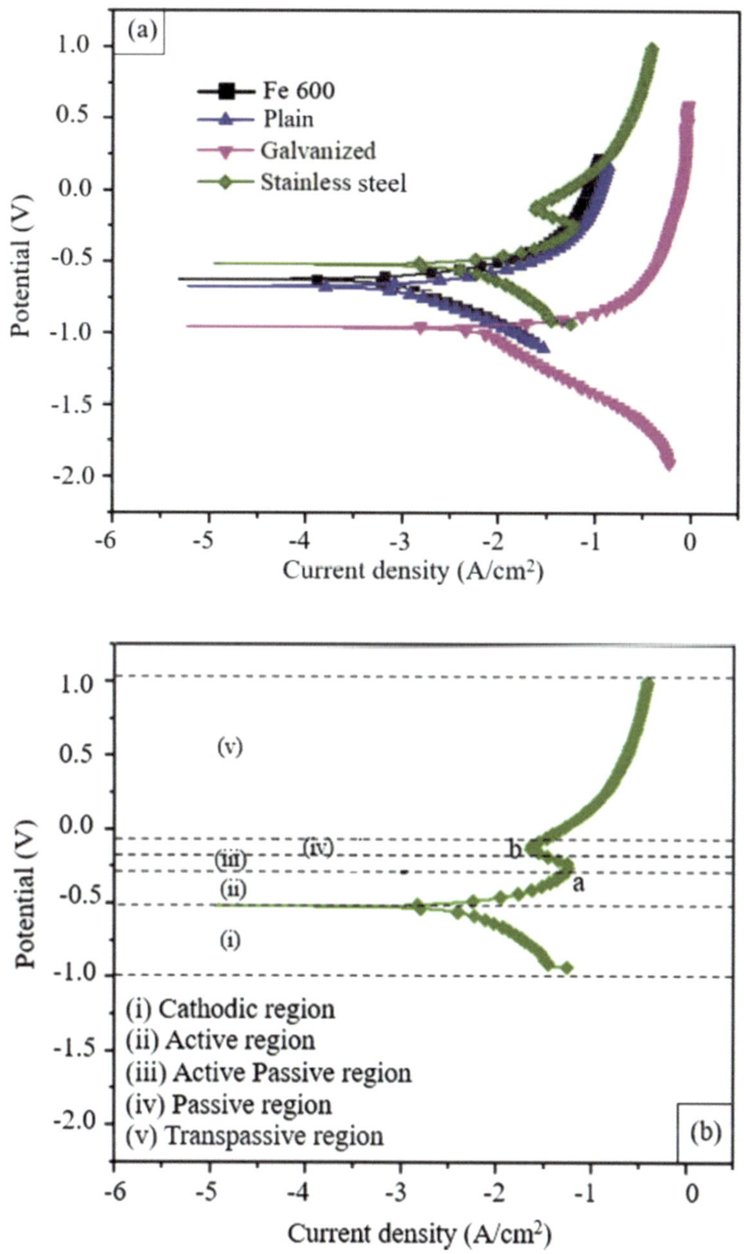

Figure 8.
(a) Tafel plots of different rebars in 1% HCl solution and (b) Tafel plot of stainless steel rebar in 1% HCl solution.

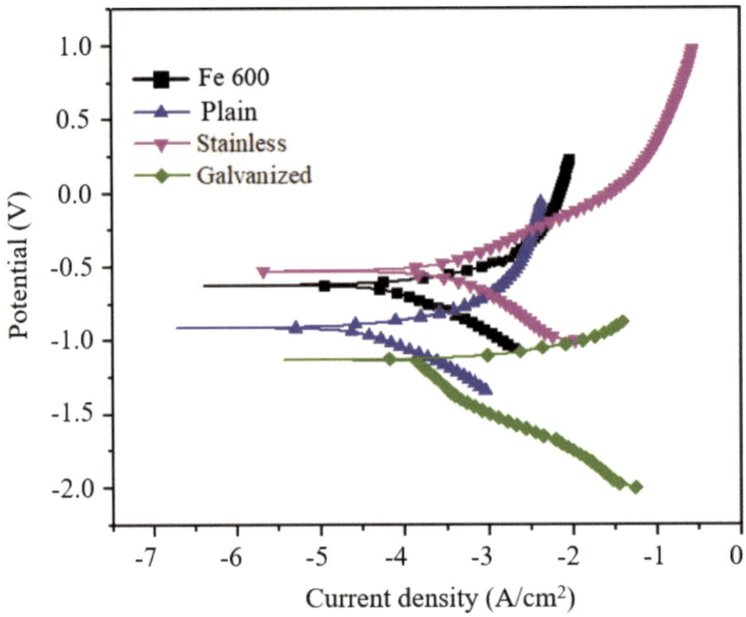

Figure 9.
Tafel plots of the rebars in 3.5% NaCl solution.

indeterminate (trans passive region), depending on the potential difference between the steel and reference electrode. **Figure 8(b)** shows that the polarization process is divided into the different zones for stainless steel at 1% HCl: (i) Cathodic region: From the corrosion potential of −0.969 V to −0.519 V, current density gradually decreases in the process. (ii) Activation region: Between −0.519 V and −0.275 V, the current density gradually increases in this region as rate of metal dissolution is higher than the rate of metal oxide formation and thereby accelerates the corrosion. (iii) Active passive region: From −0.275 V to −0.176 V, the current density starts decreasing here as the rate of metal oxide formation becomes greater than the rate of metal dissolution and metal slowly starts moving towards the passive region. (iv) Passive region: From −0.176 V to −0.058 V, metal undergoes complete passivation due to formation of chromium oxide layers and the corrosion current density becomes constant and corrosion resistance behaviour increases. (v) Trans-passive region: Above −0.058 V to 1.021 V, the passivation film is punctured due to various anodic processes, such as evolution of oxygen from water and/or the chromium oxide dissolution through Cr^{3+} to the Cr^{6+} when metal suffers high anodic polarizations; thereby, the current density increases in this region and corrosion resistance again starts decreasing. It is evident from **Figure 8(b)** that the value of current density is decreasing from point a to b, indicating greater ease of passivation by forming a stable passive film, which is similar to earlier observations [49–51]. In this connection, it is pertinent to mention here that Cr plays a major role during the transition to passive states. It has been reported earlier that increasing the Cr percentage can significantly decrease the corrosion current density values [49]. However, the analysis of the corrosion performance of different stainless steel rebars in strongly acidic environments for long-time exposure is not only limited to the effect of the elements that increase their nobility like Ni but also the effect of elements that promotes their passivation, like Cr [49]. Earlier studies have shown that the addition of both chromium and nickel to iron remarkably increases the ease of passivation [49]. It is noteworthy to mention here that when the corrosion medium changes from 1% HCl to 3.5% NaCl (**Figure 8** *vis à vis* **Figure 9**), the anodic current density decreases as depicted by a shift of the curve along the X-axis (current density) from a value of −3 A/cm^2 to values of ((−4) to (−5)) A/cm^2 due to neutral nature of 3.5% NaCl solution and thereby showing enhanced corrosion resistance of the different types of rebars in 3.5% NaCl solution. It is evident from **Figure 9**, that the current density of Fe 600 rebar is more than the plain rebar; that is, the corrosion rate of Fe 600 rebar is greater than plain rebar. Therefore, it can be assumed from **Figures 8** and **9** that although the corrosion resistance of Fe 600 rebar is higher than plain rebar in the acid medium and it is reversed in the case of seawater. It has been also observed that the corrosion resistance of plain rebar is more than galvanized rebar because the zinc coating at the surface of galvanized rebar gets damaged due to penetration of chloride ions. In this context, it cannot be overruled that the galvanized rebar undergone through thermomechanical treatments leads to the formation of a finer rim of tempered martensite, which increases the grain boundary area, thereby leading to a higher solution attack at the grain boundary and reducing the corrosion resistance of galvanized rebars.

The corresponding polarization parameters determined by extrapolation methods from the polarization curves (**Figures 8** and **9**) are summarized in **Tables 5** and **6**. From **Table 5**, it can be seen that the corrosion potential values were more positive for Fe 600, followed by plain and stainless steel rebars in 1% HCl solution, whereas in the case of 3.5% NaCl solution, more positive corrosion potential values have been observed for stainless steel followed by plain and Fe 600 rebars as evident in **Table 6**. It is well established that the higher values of corrosion potentials lead to enhancement in the corrosion resistance properties [50, 52]. From **Tables 5** and **6**, it is noticeable that the values of i_{corr} were minimum for Fe 600 followed by plain, stainless steel and Galvanized rebars in both 1% NaCl and 3.5% HCl solutions. It is well known that higher i_{corr} values represent the higher corrosion resistance behaviour of metal. **Tables 5** and **6** also summarize the estimated values of corrosion rates that were calculated by using Eq. (2) for all the experimental rebars in both 1% HCl and 3.5% NaCl solutions. It has been observed that the

Rebars	Corrosion potential (V)	β_a (mV/dec)	β_c (mV/dec)	i_{corr} (A/cm^2)	Corrosion rate (mm/year)
Fe 600	−0.565	83.0	−279.4	0.0006219	7.225
Galvanized	−1.056	400	−2.1	0.0089603	104.09
Stainless steel	−0.588	126.9	−276.6	0.0012725	14.847
Plain	−0.572	84	−222.5	0.0007222	8.390

Table 5.
Tafel plot data of five different rebars in 1% HCl solution.

Rebars	Corrosion potential (V)	β_a (mV/dec)	β_c (mV/dec)	icorr (A/cm^2)	Corrosion rate (mm/year)
Fe 600	−0.779	356.1	−125.4	0.0000801	0.930
Galvanized	−1.131	50.3	−2438.2	0.0002522	2.929
Stainless steel	−0.527	253.3	−273.2	0.000133	1.545
Plain	−0.666	108	−262.9	0.0000744	0.864

Table 6.
Tafel plot data of five different rebars in 3.5% NaCl solution.

corrosion rate of all the experimental rebars is almost the same and very small in 3.5% NaCl as compared with 1% HCl solution. However, when the solution is 1% HCl, mm/y values obtained from **Table 5** show an increase for all the experimental rebar samples, which is well expected because the acid solutions generally possess a higher corrosive environment due to their acidic nature and pH lowering properties along with the presence of water and air, which is not seen in case of 3.5% NaCl solution due to its neutral nature [34, 53]. It is well known that a lower value of corrosion rate depicts higher corrosion resistance behaviour of the samples. According to Faraday's law, the corrosion rate can be calculated by using the corrosion current density as follows [49, 52].

$$\text{Corrosion Rate}\left(\frac{\text{mm}}{\text{year}}\right) = \left(3.16 \times 10^8 i_{corr} M\right)/(zF\rho) \tag{2}$$

where i_{corr} is the corrosion current density in (A/cm^2), M is the molar mass of steel in (g/mol), F is the Faraday's constant (96,500 C/mol), z is the number of electrons transferred for each metal atom and ρ is the metal density (g/cm^3).where β_a = Anodic Tafel slope (mV/dec) and β_c = Cathodic Tafel slope (mV/dec).

Therefore, it is evident from these experimental results that the estimated values of corrosion rate were the lowest for Fe 600 followed by plain, stainless steel and galvanized rebars in acid solutions, but in the case of 3.5% NaCl solution, these values were significantly reduced for all the rebars, which is an indication of better corrosion resistance properties under marine environment.

3.4.2 Electrochemical impedance spectroscopy (EIS)

Figures 10 and **11** show the open-loop potential stabilities in the Nyquist plots of the experimental rebars in 1% HCl and 3.5% NaCl solutions, respectively. To compare with the results obtained from the polarization test (**Figures 8** and **9**) and also to get a better correlation, EIS studies were further performed. It is evident from

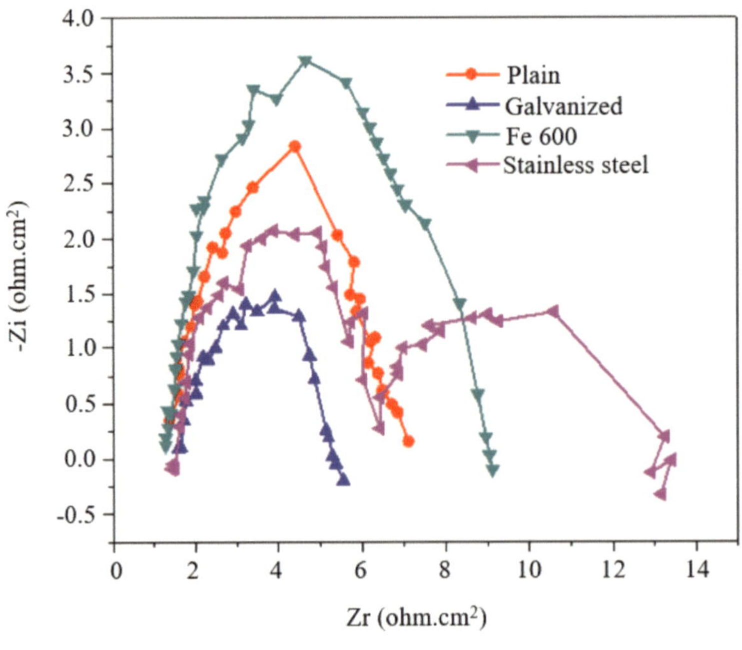

Figure 10.
Nyquist plots of the experimental rebars in 1% HCl solution.

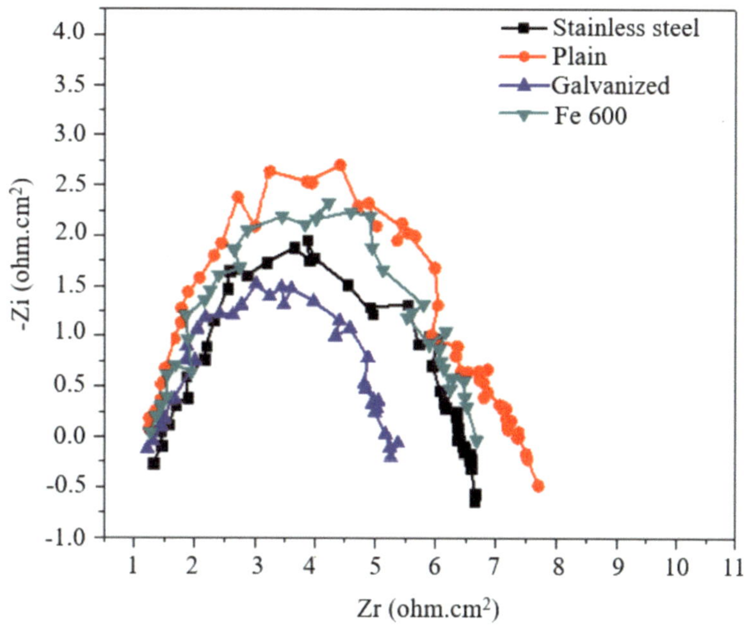

Figure 11.
Nyquist plots of the rebars in 3.5% NaCl solution (seawater).

Figure 10 that except stainless steel rebar, all other rebars are showing only one semi-circular loop in the electrochemical impedance spectra, thereby indicating a capacitive semicircle in the high-medium frequency range. The occurrence of two semicircles in the electrochemical impedance spectra for stainless steel in 1% HCl is due to passivation of the rebar surface, thereby indicating a capacitive semicircle and an inductive loop in the high-medium and low-frequency ranges, respectively [52]. The capacitive semicircle represents the active state of the interface between the rebar and acid solution when the carbon steel is exposed to the corrosive solutions [52, 54]. It is well known that stainless steel will form a trans-passive zone after the damage of the passive layer, which is also evident in **Figure 8(b)**. This is due to the evolution of oxygen from water and/or the chromium oxide dissolution through Cr^{3+} to the Cr^{6+} when metal suffers high anodic polarization [49–51]. For this reason, in the Nyquist plot two loops are visible (one is bigger and the other one is smaller). However, it cannot be overruled that the open-loop diameters in **Figure 11** were maximum for the plain rebar sample, followed by Fe 600 rebar, stainless steel, and galvanized rebar. In this connection, it is noteworthy to mention here that the capacitive arcs overlap in the Nyquist plots (**Figures 10** and **11**) since concrete is a heterogeneous material and many intermixed interfacial regions influence the impedance spectra. It has been already reported that these imperfect interfaces and electrode surface roughness require a constant phase element (CPE) instead of a pure capacitor to accurately model an equivalent circuit [55].

Tables 7 and **8** summarize the values of the resistance of the electrolyte solution (R_s), the resistance of reinforced concrete (R_f) and CPE for all the rebars. It is well established that polarization resistance ($R_p = R_f + R_{ct}$) is an indicator to study the corrosion resistance behaviour of carbon steel in corrosive environments [52]. It cannot be overruled that except for stainless steel all the plots in **Figures 10** and **11** exhibit only single-loop curves for all the other rebars, which generally do not involve the charge transfer resistance (R_{ct}). Thus, the higher values of R_f reveal a higher value of corrosion resistance for the sample [52, 56]. Furthermore, the arc diameter in the Nyquist diagrams can be considered as R_f and the reduction of the

Rebar specimen	R_s (ohm.cm^2)	R_f (ohm.cm^2)	CPE (μF.cm^{-2})
Fe 600	0.23956	13.435	105.58
Galvanized	1.6650	3.4911	2.7809
Stainless steel	2.0907	3.6110	2.8115
Plain	2.0284	10.445	30.402

Table 7.
EIS plot data of five different rebars in 1% HCl solution.

Rebar specimen	R_s (ohm.cm^2)	R_f (ohm.cm^2)	CPE (μF.cm^{-2})
Fe 600	1.5403	4.7730	2.6929
Galvanized	1.4575	3.9449	1.5659
Stainless steel	1.8498	4.6947	2.6487
Plain	1.7141	5.5280	2.8562

Table 8.
EIS plot data of five different rebars in 3.5% NaCl solution.

arc diameter reveals a decrease in R_f values [52]. It is evident from **Table 7** that the R_f values were maximum for Fe 600, followed by plain, stainless steel, and then galvanized rebars in 1% HCl solution, whereas in 3.5% NaCl solution (**Table 8**), it is showing maximum values for plain rebar followed by Fe 600 and stainless steel rebars with a slight variation (≈0.07) and then galvanized rebar, which is also in agreement with **Figures 10** and **11**. Therefore, all the experimental evidence obtained from the polarization and EIS studies under acidic and seawater solutions are in good agreement and the corrosion resistance of all the experimental rebars is higher in 3.5% NaCl than 1% HCl solutions.

4. Conclusions

The important conclusions drawn from this study are as follows:

1. Microstructures of all the TMT rebars consist of an outer martensitic rim with an intermediate narrow bainitic transition zone followed by a ferrite-pearlite inner core, whereas stainless steel and plain rebars show equiaxed ferrite grain and ferrite-pearlite microstructure throughout the sample with no transition zone between core and rim due to the absence of TMT.

2. Fe 600 and galvanized rebars exhibit classical U-shaped hardness profiles, whereas stainless steel and plain rebars reveal a different variation at the edges when compared with other rebars and the absence of the classical U-shaped hardness profiles.

3. Stainless steel rebar shows the highest combination of strength (≈755 MPa) and ductility (27%) among all the rebars followed by Fe 600 rebar (≈660 MPa) in terms of strength. Plain rebar although showing lower values of strength (≈470 MPa) shows the maximum amount of ductility (≈29%). Therefore, it is expected that all the experimental rebars will show improved performance in the seismic zones.

4. In the Tafel plot and Nyquist plot, stainless steel shows a different property in 1% HCl solution than the others. This is due to the formation of the trans-passive zone *via* the evolution of oxygen from water and/or the chromium oxide dissolution through Cr^{3+} to the Cr^{6+} when metal suffers high anodic polarization.

5. In the Nyquist plot, a higher R_p value indicates higher corrosion resistance. Also, the larger the radius of the capacitive loop, the greater will be the charge transfer resistance, that is, corrosion resistance. It is also true that greater current density (i_{corr}) signifies a higher corrosion rate in the Tafel plot. Therefore, it can be concluded that in both Tafel and Nyquist studies, Fe 600 rebar has revealed maximum corrosion resistance followed by plain, stainless steel and galvanized rebars in 1% HCl solution, whereas in 3.5% NaCl solution, plain rebar has shown maximum corrosion resistance and then followed by Fe 600 and stainless steel rebars with a minor variation and then galvanized rebar.

Acknowledgements

The authors are grateful to Dr. Anil Kumar Kar for supplying the plain rebar required in this investigation.

Author details

Indrajit Dey[1], Pallabi Manna[1], Muralidhar Yadav[1], Nisith Kumar Tewary[1], Jayanta Kumar Saha[2] and Swarup Kumar Ghosh[1*]

1 Department of Metallurgy and Materials Engineering, Indian Institute of Engineering Science and Technology, Shibpur, Howrah, West Bengal, India

2 Institute for Steel Development and Growth, Kolkata, West Bengal, India

*Address all correspondence to: skghosh@metal.iiests.ac.in

References

[1] Li X, Lo KH, Kwok CT, Sun YF, Lai KK. Post-fire mechanical and corrosion properties of duplex stainless steel: Comparison with ordinary reinforcing-bar steel. Construction and Building Materials. 2018;**174**:150-158. DOI: 10.1016/j.conbuildmat.2018.04.110

[2] Coccia S, Imperatore S, Rinaldi Z. Influence of corrosion on the bond strength of steel rebars in concrete. Materials and Structures. 2016;**49**: 537-551. DOI: 10.1617/s11527-014-0518-x

[3] Apostolopoulos CA, Papadakis VG. Consequences of steel corrosion on the ductility properties of reinforcement bar. Construction and Building Materials. 2008;**22**(12):2316-2324. DOI: 10.1016/j.conbuildmat.2007.10.006

[4] Tae SH, Kyung JW, Ujiro T, Shin SW. Service life estimation of reinforced concrete structures made using Cr-bearing rebars in microcell corrosion environments. ISIJ International. 2007;**47**(9):1315-1323. DOI: 10.2355/isijinternational.47.1315

[5] Baddoo NR. Stainless steel in construction: A review of research, applications, challenges and opportunities. Journal of Constructional Steel Research. 2008;**64**(11):1199-1206. DOI: 10.1016/j.jcsr.2008.07.011

[6] Serdar M, Zulj LV, Bjegovic D. Long-term corrosion behaviour of stainless reinforcing steel in mortar exposed to chloride environment. Corrosion Science. 2013;**69**:149-157. DOI: 10.1016/j.corsci.2012.11.035

[7] Moser RD, Singh PM, Kahn LF, Kurtis KE. Chloride-induced corrosion resistance of high-strength stainless steels in simulated alkaline and carbonated concrete pore solutions. Corrosion Science. 2012;**57**:241-253. DOI: 10.1016/j.corsci.2011.12.012

[8] Alvarez SM, Bautista A, Velasco F. Corrosion behaviour of corrugated lean duplex stainless steels in simulated concrete pore solutions. Corrosion Science. 2011;**53**:1748-1755. DOI: 10.1016/j.corsci.2011.01.050

[9] Bautista A, Blanco G, Velasco F, Gutierrez A, Soriano L, Palomares FJ, et al. Changes in the passive layer of corrugated austenitic stainless steel of low nickel content due to exposure to simulated pore solutions. Corrosion Science. 2009;**51**:785-792. DOI: 10.1016/j.corsci.2009.01.012

[10] Blanco G, Bautista A, Takenouti H. EIS study of passivation of austenitic and duplex stainless steels reinforcements in simulated pore solutions. Cement and Concrete Composites. 2006;**28**: 212-219. DOI: 0.1016/j.cemconcomp.2006.01.01

[11] Luo H, Dong CF, Li XG, Xiao K. The electrochemical behaviour of 2205duplex stainless steel in alkaline solutions with different pH in the presence of chloride. Electrochimica Acta. 2012;**64**:211-220. DOI: 10.1016/j.electacta.2012.01.025

[12] Freire L, Catarino MA, Godinho MI, Ferreira MJ, Ferreira MGS, Simoes AMP, et al. Electrochemical and analytical investigation of passive films formed on stainless steels in alkaline media. Cement and Concrete Composites. 2012;**34**: 1075-1081. DOI: 10.1016/j.cemconcomp.2012.06.002

[13] Paul SK, Rana PK, Das D, Chandra S, Kundu S. High and low cycle fatigue performance comparison between micro alloyed and TMT rebar. Construction and Building Materials. 2014;**54**:170-179. DOI: 10.1016/j.conbuildmat.2013.12.061

[14] Rasouli D, Khameneh S, Akbarzadeh A, Daneshi GH. Effect of

cooling rate on the microstructure and mechanical properties of micro alloyed forging steel. Journal of Materials Processing Technology. 2008;**206**: 92-98. DOI: 10.1016/j.jmatprotec.2007.12.006

[15] Opiela M. Thermo-mechanical treatment of the C-Mn steel with Nb, Ti, V and microadditions. Archives of Materials Science and Engineering. 2007;**28**(6):377-380

[16] Ghosh A, Ghosh M. Tensile and impact behaviour of thermo mechanically treated and micro-alloyed medium carbon steel bar. Construction and Building Materials. 2018;**192**: 657-670. DOI: 10.1016/j.conbuildmat.2018.10.098

[17] Ceschini L, Marconi A, Martini C, Morri A, Schino ADI. Tensile and impact behaviour of a microalloyed medium carbon steel: Effect of the cooling condition and corresponding microstructure. Materials and Design. 2013;**45**:171-178. DOI: 10.1016/j.matdes.2012.08.063

[18] Khan MR, Khan MM. Comparative study of the micro-structure of quenched self-tempered and ordinary steel. Journal of Pakistan Materials Society. 2008;**2**(2):82-86

[19] Rocha M, Michel S, Bruhwiler E, Nussbaumer A. Very high cycle fatigue tests of quenched and self-tempered steel reinforcement bars. Materials and Structures. 2016;**49**:1723-1732. DOI: 10.1617/s11527-015-0607-5

[20] Bertolini L, Elsener B, Pedeferri P, Redaelli E, Polder RB. Corrosion of Steel in Concrete: Prevention, Diagnosis, Repair. 2nd ed. Weinheim: Wiley-VCH; 2013

[21] Venkatesan P, Palaniswamy N, Rajagopal K. Corrosion performance of coated reinforcing bars embedded in concrete and exposed to natural marine environment. Progress in Organic Coatings. 2006;**56**:8-12. DOI: 10.1016/j.porgcoat.2006.01.011

[22] Erdogdu S, Bremner TW, Kondratova IL. Accelerated testing of plain and epoxy-coated reinforcement in simulated seawater and chloride solutions. Cement and Concrete Research. 2001;**31**:861-867

[23] Nandi SK, Tewary NK, Saha JK, Ghosh SK. Microstructure, mechanical properties and corrosion performance of a few TMT rebars. Corrosion Engineering Science and Technology. 2016;**51**(7):476-488. DOI: 10.1080/1478422X.2016.1141744

[24] Yu L, Francois R, Dang VH, L'Hostis V, Gagne R. Development of chloride-induced corrosion in pre-cracked RC beams under sustained loading: Effect of load-induced cracks, concrete cover, and exposure conditions. Cement and Concrete Research. 2015;**67**: 246-258. DOI: 10.1016/j.cemconres.2014.10.007

[25] Apostolopoulos CA, Demis S, Papadakis VG. Chloride induced corrosion of steel reinforcement-mechanical performance and pit depth analysis. Construction and Building Materials. 2013;**38**:139-146. DOI: 10.1016/j.conbuildmat.2012.07.087

[26] Ye CQ, Hu RG, Dong SG, Zhang XJ, Hou RQ, Du RG, et al. EIS analysis on chloride-induced corrosion behaviour of reinforcement steel in simulated carbonated concrete pore solutions. Journal of Electroanalytical Chemistry. 2013;**688**:275-281. DOI: 10.1016/j.jelechem.2012.09.012

[27] Song HW, Saraswathy V, Muralidharan S, Lee CH, Thangavel K. Corrosion performance of steel in composite concrete system admixed with chloride and various alkaline nitrites. Corrosion Engineering Science and Technology. 2009;**44**(6):408-415. DOI: 10.1179/174327809X397848

[28] Rivera-CorraL JO, Fajardo G, Arliguie G, Orozco-Cruz R, Deby F, Valdez P. Corrosion behavior of steel reinforcement bars embedded in concrete exposed to chlorides: Effect of surface finish. Construction and Building Materials. 2017;**147**:815-826. DOI: 10.1016/j.conbuildmat.2017.04.186

[29] Dong SG, Zhao B, Lin CJ, Du RG, Hu RG, Zhang GX. Corrosion behavior of epoxy/zinc duplex coated rebar embedded in concrete in ocean environment. Construction and Building Materials. 2012;**28**:72-78. DOI: 10.1016/j.conbuildmat.2011.08.026

[30] Kar AK. Rebars for Durable Concrete Construction: Points to Ponder. Design of Cities and Buildings-Sustainability and Resilience in the Built Environment. London: IntechOpen; 2021. pp. 13-35

[31] Kamimura T, Hara S, Miyuki H, Yamashita M, Uchida H. Composition and protective ability of rust layers formed on weathering steel exposed to various environments. Corrosion Science. 2006;**48**:2799-2812. DOI: 10.1016/j.corsci.2005.10.004

[32] Loto CA, Loto RT. Electrochemical corrosion resistance evaluation of ferritic stainless steel in HCl. International Journal of Electrochemical Science. 2012;**7**:11011-11022

[33] Panigrahi BK, Srikanth S, Sahoo G. Effect of alloying elements on tensile properties, microstructure, and corrosion resistance of reinforcing bar steel. Journal of Materials Engineering and Performance. 2009;**18**:1102-1108. DOI: 10.1007/s11665-008-9336-z

[34] Tewary NK, Kundu A, Nandi R, Saha JK, Ghosh SK. Microstructural characterization and corrosion performance of old railway girder bridge steel and modern weathering structural steel. Corrosion Science. 2016;**113**:57-63. DOI: 10.1016/j.corsci.2016.10.004

[35] Wu Z, Troparevsky MC, Gao YF, Morris JR, Stocks GM, Bei H. Phase stability, physical properties and strengthening mechanisms of concentrated solid solution alloys. Current Opinion in Solid State and Materials Science. 2017;**21**:267-284. DOI: 10.1016/j.cossms.2017.07.001

[36] Hasan MA, Yan K, Lim S, Akiyama M, Frangopol DM. LCC-based identification of geographical locations suitable for using stainless steel rebars in reinforced concrete girder bridges. Structure and Infrastructure. Engineering. 2020;**16**:1-27. DOI: 10.1080/15732479.2019.1703758

[37] Vlack LHV. Elements of Material Science and Engineering. Vol. 4. Massachusetts: Adison-Wesley Reading; 1987. pp. 31-32

[38] Song HW, Saraswathy V. Corrosion monitoring of reinforced concrete structures–a review. International Journal of Electrochemical Science. 2007;**2**:1-28

[39] Panigrahi BK, Jain SK. Impact performance of high strength low alloy TMT reinforcement ribbed bar. Bulletin of Materials Science. 2002;**25**(4): 319-324

[40] Batusia A, Pomares JC, Gonzalez MN, Velasco F. Influence of the microstructure of TMT reinforcing bars on their corrosion behaviour in concrete with chlorides. Construction and Building Materials. 2019;**229**: 116899. DOI: 10.1016/j.conbuildmat.2019.116899

[41] Paul SK, Majumdar S, Kundu S. Low cycle fatigue behavior of thermo-mechanically treated Rebar. Materials

and Design. 2014;**58**:402-411. DOI: 10.1016/j.matdes.2014.01.079

[42] Khalifa H, Megahed GM, Hamouda RM, Taha MA. Experimental investigation and simulation of structure and tensile properties of Tempcore treated rebar. Journal of Materials Processing Technology. 2016;**230**: 244-253. DOI: 10.1016/j.jmatprotec. 2015.11.023

[43] Sk MB, Khan AK, Lenka S, Syed B, Chakraborty J, Chakrabarti D, et al. Effect of microstructure and texture on the impact transition behaviour of thermo-mechanically treated reinforcement steel bars. Materials and Design. 2016;**90**:1136-1150. DOI: 10.1016/j.matdes.2015.11.053

[44] Ahmed E, Ibrahim S, Galal M, Elnekhaily SN, Allam T. Microstructure and mechanical properties of V-alloyed rebars subjected to Tempcore process. Metals. 2021;**11**(246):1-17. DOI: 10.3390/met11020246

[45] Osarolube E, Owate IO, Oforka NC. Corrosion behaviour of mild and high carbon steels in various acidic medium. Scientific Research and Essay. 2008; **3**(6):224-228

[46] Odebiyi OS, Adedayo SM, Tunji LA, Onuorah MO. A review of weldability of carbon steel in arc-based welding processes. Cogent Engineering. 2019; **6**(1):1609180. DOI: 10.1080/ 23311916.2019.1609180

[47] Rajkumar C. Increasing the Yield Strength of Niobium Microalloyed Reinforcing Bar. Research Report (master of science). Johannesburg: University of Witwatersrand; 2008. pp. 14-90

[48] Zou G, Shi W, Xiang S, Ji X, Ma G, Ballinger RG. Corrosion behavior of 904L austenitic stainless steel in hydrofluoric acid. The Royal Society of Chemistry Advances. 2018;**8**:2811-2817. DOI: 10.1039/C7RA12453H

[49] Bellezze T, Giuliani G, Roventi G. Study of stainless steels corrosion in strong acid mixture. Part 1: Cyclic potentiodynamic polarization curves examined by means of analytical method. Corrosion Science. 2018;**130**: 113-125. DOI: 10.1016/j. corsci.2017.10.012

[50] Luo H, Su H, Dong C, Xiao K, Li X. Electrochemical and passivation behavior investigation of ferritic stainless steel in simulated concrete pore media. Data in Brief. 2015;**5**:171-178. DOI: 10.1016/j.dib.2015.08.016

[51] Invernizzi AJ, Sivieri E, Trasatti SP. Corrosion behaviour of duplex stainless steels in organic acid aqueous solutions. Materials Science and Engineering A. 2008;**485**:234-242. DOI: 10.1016/j. msea.2007.08.036

[52] Zhou H, Wang Y, Ma T. Effect of silicon addition on corrosion behaviour of carbon steel rebar in sulphuric acid environment. International Journal of Electrochemical Science. 2020;**15**: 3003-3012. DOI: 10.20964/2020.04.22

[53] Bar HN, Sivaprasad S, Narasaiah N, Paul SK, Sen BN, Chandra S. Low cycle and ratchetting fatigue behaviour of high UTS/ YS ratio reinforcing steel bars. Journal of Materials Engineering and Performance. 2013;**22**:1701-1707. DOI: 10.1007/s11665-013-0470-x

[54] Husairi F, Rouhi J, Eswar K, Zainurul AZ, Rusop M, Abdullah S. Electrochemical impedance spectroscopy analysis of porous silicon prepared by photo-electrochemical etching: Current density effect. Applied Physics A. 2014;**116**:2119-2124. DOI: 10.1007/s00339-014-8416-1

[55] Sohail MG, Kahraman R, Alnuaimi NA, Gencturk B, Alnahhal W,

Dawood M, et al. Electrochemical behavior of mild and corrosion resistant concrete reinforcing steel. Construction and Building Materials. 2020;**232**: 117205. DOI: 10.1016/j.conbuildmat.2019.117205

[56] Rouhi J, Mahmud S, Hutagalung SD, Kakooei S. Journal of Micro/Nanolithography, MEMS, and MOEMS. 2011;**10**:043002. DOI: 10.1117/1.3643480